AF882338

U.S. ARMY SURVIVAL MANUAL

U.S. ARMY SURVIVAL MANUAL

Dorset Press
New York

This edition published by Dorset Press,
a division of Marboro Books Corp.,
by arrangement with Platinum Press, Inc.
1994 Dorset Press

ISBN 1-56619-022-3

U.S. ARMY SURVIVAL MANUAL

Table of Contents

Preface

No one knows survival better than the U.S. Army, so this exceptional field guide is the most authoritative of its kind. Originally commissioned by the Department of the Army to train its special forces in all-climate, all-terrain survival tactics, this newly issued edition is meant to serve as "a civilian's best guide for toughing it...anyplace in the world."

A must for campers, hikers, explorers, pilots, and others whose vocation or avocations require familiarity with the wilderness or out-of-doors, this excellent manual describes and clearly illustrates the techniques of survival medicine, tool making, food and water procurement, shelter building, direction finding, signaling, and many others that could mean the difference between life and death in remote areas. The book will also stimulate the juices of armchair adventurers.

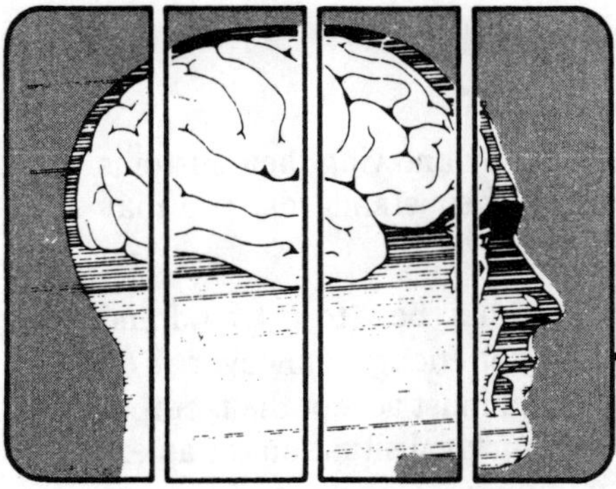

Chapter 1

The Will to Survive

Many survival case histories show that stubborn, strong willpower can conquer many obstacles. One case history tells of a man stranded in the desert for eight days without food and water; he had no survival training, and he did nothing right. But he wanted to survive, and through sheer willpower, he did survive.

With training, equipment, and the ***will to survive,*** *you will find you can overcome any obstacle you may face. You will survive.*

Preparing for Survival

Two things that you can *do now* to help you prepare are (1) train for survival in different environments and (2) learn about the area where you are going.

Learn how to find and get food and water, how to use natural environmental features for shelter, how to build shelter, how to determine direction, and how and when to travel through different types of terrain.

You should learn how to maintain your health, how to avoid environmental hazards, and how to doctor yourself.

Most important, you should learn that rest can be more valuable than speed. Whether you are struggling through jungle undergrowth, facing a dwindling water supply in the desert, or making your way across arctic ice, you should plan and make your way carefully; do not dash blindly on.

You should learn about the natives in the area(s) where you expect to go. This knowledge and common sense will enable you to make contact with them. If you reach an area where the natives are not allied with the enemy, you will have little to fear and much to gain by thoughtful contact with them. By knowing their customs in advance and by being courteous by their standards, you may be able to get their help.

Learning about the different environments, how to get food and water, how to take care of yourself is not enough, however. You must have the right attitude. That is, you must accept the fact that as a soldier you may find yourself in a survival situation for an extended period of time, alone, with minimum equipment. You must understand that this situation could come about without warning, and you must be prepared.

If you have the opportunity, go through a survival school in which you can train and practice your skills before they really count. The training you receive will give you confidence in yourself. A few hours spent in applying what you learn—finding that it works, finding that you can accomplish things that previously seemed impossible—will remove any doubt about your personal ability and stamina.

Survival Stresses

You must understand the emotional states associated with survival just as you must understand survival conditions and equipment. In a survival situation, you (and your companions, if any) are the most important element in determining your success or failure.

You have probably never given much thought to—

- How do I react to various situations?
- What do the various signs, feelings, expressions, and reactions in me (and in others) mean?
- What are my tolerances to different physical and mental stresses?
- How can I maintain and use my abilities effectively to perform and to control myself?
- How can I influence my companions in a way that will help them and me?

Yet, knowing these things—knowing "thyself"—is extremely important in a survival situation and bears directly on how well you cope with serious stresses: fear and anxiety; pain, injury, and illness; cold and heat; thirst; hunger; fatigue; sleep deprivation; boredom; and loneliness and isolation.

Your body has reaction mechanisms that generally help you to adapt to stress. For example, when you are scared or angry, these mechanisms cause changes within your body that tend to increase your alertness and to provide you with extra energy. These reaction mechanisms, however, can betray you under survival conditions. For instance, your "normal" reaction to hunger (a stress) might cause you to cast reason aside and forage for food in enemy territory, exposing yourself to capture.

Two of the gravest general dangers to survival are the desire for comfort and a passive outlook. You must recognize that these dangers represent attitudes—attitudes that follow lines of least resistance, that overrule your effort or desire to cope with stress, that make your primary concern the immediate situation rather than the overall problem of survival.

To overcome the first danger—*the desire for comfort*—you need to change the way you think of comfort. And the key to changing is reasoning: You compare your present discomfort with the discomfort you will face if captured. Your present discomfort is a temporary problem; as a prisoner your discomfort would probably continue indefinitely and be more intense. Knowing how much discomfort you can take and understanding your demand for comfort will help you to carry on. Comfort is not essential!

To avoid the second danger—*the passive outlook*—you should know what can bring it on.

Some physical conditions contribute to the passive outlook. They include exhaustion due to prolonged exposure to cold, excessive loss of body fluids (dehydration), excessive fatigue, weakness, and illness. You can avoid these conditions by proper planning and sound decisions.

Lack of the will to keep trying can also result in a passive outlook. Lethargy, mental numbness, and indifference creep in slowly, but they can suddenly take over and leave you helpless.

Recognizing the onset of a passive outlook in a companion is important. The first signs are an air of resignation, quietness, lack of communication, loss of appetite, and withdrawal from the group. The best way to deal with such an outlook is to stop or to counter the physical and mental stresses that produce it.

Following are the enemies of survival. They are common physical and mental stresses that will impact on your ability to cope with a survival situation.

Pain. Pain is your body's way of telling you that you have an injury. Pain, in itself, is not harmful, but it does make you uncomfortable. You may not notice pain if your mind is concentrated on other matters. But if you let it, pain can get the best of you. It can weaken your will to survive. You can tolerate pain, however, if you—

- Understand its source and nature.
- Recognize it as something to be tolerated.
- Concentrate on things you need to do (think, plan, keep busy).
- Take pride in your ability to take it.

Cold. Being cold lowers your efficiency, your ability to think, and your will to do anything but try to get warm. It slows down the flow of blood so that you get sleepy. These effects of cold are dangerous. You should immediately seek shelter and build a fire to warm yourself before the cold weakens your will to survive.

Heat. Weakness is the principal symptom of unaccustomed exposure to heat. You can, however, become accustomed to high temperatures. It may take 2 to 6 days for your circulation, your breathing, your heart action, and your sweat glands to get adjusted to a hot climate. When you are in direct sunlight in hot climates keep your head covered. If your situation allows, do not exert yourself during the hottest hours of the day.

Thirst. Two of the most serious problems of survival are thirst and dehydration. Thirst, even though not extreme, can dull your mind. Drink plenty of water when there is an ample supply available, particularly when eating. If your water supply is low, cut down your food intake. Your body must use extra water to carry off wastes from food; this causes the body to become dehydrated more quickly. You can reverse almost any stage of dehydration simply by drinking enough water.

Hunger. Unrelieved hunger will affect attitude, morale, and will to survive. If this condition does not change, undernourishment/starvation will bring on the following symptoms: loss of weight, weakness, dizziness and blackouts on standing up suddenly, slowed heart rate, increased sensitivity to cold, and increased thirst. In many areas you can find a lot of edible material that you may not regard as food (Chapter 6). To survive, you must overcome your food dislikes and prejudices and eat whatever edible items are available.

Fatigue. Being tired can reduce your mental ability, make you careless, and give you an I-don't-care feeling. Overexertion causes fatigue; but so does mental outlook: hopelessness, lack of a goal, dissatisfaction, frustration, and boredom. Rest, of course, is the primary means of overcoming fatigue caused by overexertion. You can tell when you have reached the state of tiredness that affects your mental and physical ability. Often you can summon the strength to go on if you recognize the dangers of a situation. Fatigue caused by mental outlook can sometimes be overcome by a change of activity, by mild exercise, or by conversation with others.

Boredom. Repetitiveness and uniformity are two sources of boredom. And with boredom comes lack of interest, feelings of strain, and anxiety or depression, especially when you can see no relief and are frustrated. To overcome boredom, you must keep your goal—survival—in mind and realize how the tasks you must perform fit into your overall survival plan.

Loneliness. Being isolated under trying and threatening conditions will often bring on another stress—a feeling of loneliness—which can lead to feelings of helplessness and despair. Self-sufficiency plays a major role in overcoming these feelings.

You have been developing self-sufficiency all your life. You learned how to do things by yourself, how to be comfortable while alone, how to accept new surroundings, and how to cope with new problems. Added to that, your military training has provided you with additional knowledge on problems and conditions you can expect. You are prepared to be active, to plan, and to think with a deliberate purpose. These are countermeasures to loneliness.

You can increase your self-sufficiency—your ability to function competently on your own—with practice. You have opportunities to do so each day of your life: Make your own decisions and rely on yourself; explore new situations and solve problems. You must learn to accept the reality of a new situation or of an emergency and then take suitable action. This is one of the most important psychological requirements for survival. Do not sit down and worry. Stay busy!

Attitude

Having a "survival attitude" for whatever may occur is extremely important. Mental and actual rehearsal of emergency procedures prepares a person to take automatic action. Knowledge and rehearsal of survival procedures will give you a feeling of confidence—a "survival attitude"—and will prepare you for any emergency, even though you may be semiconscious at the time. A person without a positive mental attitude may panic under dire circumstances.

Personal Qualities

Personality also plays an important role in survival. In fact, it may have more bearing on survival than danger, weather, terrain, or the nature of the emergency. For example, whether a person's fatigue dulls or sharpens his mind, overcomes or intensifies his ability to take necessary survival actions, or lessens or increases his determination to carry on depends, to a large extent, more on that person than on the situation. The following personal qualities are important to survival:

- Being able to make up your mind.
- Being able to improvise.
- Being able to live with yourself.
- Being able to adapt to the situation—to make a good thing out of a bad thing.
- Remaining cool, calm, and collected.
- Hoping for the best, but preparing for the worst.
- Having patience.
- Being prepared to meet the worst that can happen.
- Being able to "figure out" other people—to understand and predict what other people will do.
- Understanding where your special fears and worries come from, and knowing what to do to control them.

Survival Actions

You can reduce and overcome the shock of being isolated behind enemy lines if you keep the key word S-U-R-V-I-V-A-L foremost in your mind. Its letters can help to guide you in your actions:

S - *Size up the situation.*

If you are in a combat situation, find a place where you can conceal yourself from the enemy—remember, security takes priority. Use your senses of hearing, smell, and sight to get a feel for the battlefield. What is the enemy doing? Advancing? Holding in place? Retreating? You will have to consider what is developing on the battlefield when you make your survival plan.

Size up your surroundings.

Determine the pattern of the area. Get a feel for what is going on around you. Every environment, whether forest, jungle, or desert, has a rhythm or pattern. This rhythm or pattern includes animal and bird noises and movements and insect sounds. It may also include enemy traffic and civilian movements.

Size up your physical condition.

The pressure of the battle you were in or the trauma of being in a survival situation may have caused you to overlook wounds you received. Check your wounds and give yourself first aid. Take care to prevent further bodily harm. For instance, in any climate, drink plenty of water to prevent dehydration; if you are in a cold or wet climate, put on additional clothing to prevent hypothermia.

Size up your equipment.

Perhaps in the heat of battle, you lost or damaged some of your equipment. Check to see what equipment you have and what condition it is in.

Now that you have sized up your situation, your surroundings, your physical condition, and your equipment, you are ready to make your survival plan. In doing so, keep in mind your basic physical needs: water, food, and shelter.

U - *Undue haste makes waste.*

You may make a wrong move when you react quickly without thinking or planning, and that move may result in your capture or death. Don't move just for the sake of taking action. Consider all aspects of your situation (size up your situation) before you make a decision and a move. If you act in haste, you may forget or lose some of your equipment, and you may become disoriented so that you don't know which way to go. Plan your moves so that you are prepared to move out quickly without endangering yourself if the enemy is near you.

R - *Remember where you are.*

Spot your location on your map and relate it to the surrounding terrain. This is a basic principle that you should always follow. If there are other persons with you, make sure that they also know their location. Always know who in your group, vehicle, or aircraft has a map and compass. If that person is killed, you will have to get the map and compass from him. Pay close attention to where you are and to where you are going. Do not rely on others present to keep track of the route. Constantly orient yourself. You should always try to determine, as a minimum, how your location relates to:

- The location of enemy units and controlled areas.
- The location of friendly units and controlled areas.
- The location of local water sources (this is especially important in the desert).
- Areas that will provide good cover and concealment.

This information will allow you to make intelligent decisions when you are in a survival/evasion situation.

V - *Vanquish fear and panic.*

The greatest enemies in a combat survival/evasion situation are fear and panic. If uncontrolled, they can destroy your ability to make an intelligent decision. They may cause you to react to your feelings and imagination rather than to your situation. They can drain your energy and thereby cause other negative emotions. Previous survival/evasion training and self-confidence (page 1-2) will enable you to vanquish fear and panic.

I - *Improvise.*

In the United States we have items available for all of our needs. And many of these items are cheap to replace when damaged. This easy-to-come easy-to-go easy-to-replace culture of ours makes it unnecessary for us to improvise. This inexperience in improvisation can be one of the greatest enemies in a survival situation. Learn to improvise. Take a tool designed for a specific purpose and see how many other uses you can make of it.

Learn to use natural things around you for different needs. An example is using a rock for a hammer. No matter how complete a survival kit you have with you, it will run out or wear out after awhile. But your imagination will not. Use it.

V - *Value living.*

All of us were born kicking and fighting to live. But we have become used to the soft life. We have become creatures of comfort. We dislike inconveniences and discomforts. So, what happens when we are faced with a survival situation with its stresses, inconveniences, and discomforts? This is when the will to live—placing a high value on living—is vital. The experience and knowledge you have gained through life and through your Army training have bearing on your will to live. Stubbornness, a refusal to give in to problems and obstacles that face you, will give you the mental and physical strength to endure.

A - *Act like the natives.*

The natives and animals of a region have adapted to their environment. To get a feel of the area, watch how the people go about their daily routine. When and what do they eat? When, where, and how do they get their food? When and where do they go for water? What time do they usually go to bed and get up? These things are important to you as an evader.

Animal life in the area can also give you clues on how to survive. Animals also require food, water, and shelter. By watching them, you can find sources of water and food.

NOTE: *Animals cannot serve as an absolute guide to what you can eat and drink.*

Keep in mind that the reaction of animals to your presence can reveal your presence to the enemy.

If you are in a friendly area, one of the best ways to gain rapport with the natives is to show interest in their tools and their ways of procuring food and water. By studying the people, you will learn to respect them, you can often make valuable friends, and, most important, you can learn how to adapt to their environment and increase your chances of survival.

L - *Live by your wits,* ***but for now,*** *learn basic skills.*

Without training in basic skills for surviving and evading on the battlefield, your chances of living through a combat survival/evasion situation are slight.

The time to learn these basic skills is NOW—not when you are headed for or are in the battle. How you decide to equip yourself prior to deployment will impact on whether or not you survive. You need to know about the environment to which you are going, and you must practice basic skills geared to that environment. For instance, if you are going to a desert, you need to know how to get water in the desert.

Practice basic survival skills during all training programs and exercises. Survival training reduces fear of the unknown and gives you self-confidence. It teaches you to *live by your wits.*

Pattern for Survival

You should develop a pattern for survival to enable you to beat the enemies of survival described on pages 1-4 and 1-5. This pattern of survival must include food, water, shelter, fire, first aid, and signals placed in order of importance. For example, in a cold environment, you would need a *fire* to get warm; a *shelter* for protection from cold, wind, and rain or snow; traps or snares to procure *food;* a means to *signal* to friendly aircraft; and *first aid* to maintain health. *If you are injured, first aid has top priority* no matter what climate you are in.

You must change your pattern of survival to meet your immediate physical needs in your environment. In the desert, for instance, water will have top priority if you are not injured.

Chapter 2

Survival Planning

*Survival planning is nothing more than realizing that something could happen that would put you in a survival situation and, with that realization, taking steps to increase your chances of survival. In other words, survival planning is **being prepared.***

Being prepared means you have survival items with you, and know how to use them. For instance, people who live in snow regions prepare their vehicles for poor road conditions. They put snow tires on their vehicles; they add extra weight in the back for traction; and they carry a shovel, salt, and a blanket. They realize they could get stuck in snow, and they are prepared. Another example of being prepared is finding the exits on an aircraft when you board it for a flight. You realize that something could happen requiring you to get out of the plane quickly, and you prepare for it. You note where the exits are located—to your front or back, to your left or right, the number of rows away from you.

You read in Chapter 1 about the will to survive. Until you face a survival situation, however, you can only "think" rather than "know" you have that will to survive. But you can train, practice, and prepare to meet any survival challenge.

Importance of Planning

The details in survival planning may seem of little importance when compared with those in military operations planning. They can become quite significant, however, if a survival situation arises. Imagine how your chances of success in evading an enemy would decrease if you were wearing new or improperly fitted boots. A person who normally is not required to walk long distances will tend to overlook the importance of proper footgear.

A small detail that is extremely important when working in an arid area is conserving water. You must make maximum use of the available water. If you have troops serving under you, for instance, you might devise a means to catch the water that overflows when they fill their canteens from 5-gallon cans. They can then use the overflow for washing and shaving.

Another small detail is preventive medicine. Have your teeth checked and your shots up to date before you go on a mission. A tooth problem in a survival situation will greatly reduce your ability to cope with other problems you may face. Failure to keep your shots up to date may mean your body is not immune to diseases that are prevalent in the area.

If your job requires that you work in a small enclosed area that limits what you can carry on your person, plan where you can put your rucksack or your load-bearing equipment. Put it where it will not prevent you from getting out quickly, yet where you can grab it quickly.

Survival Kit

Just as important as the preparations and plans mentioned above is preparing and carrying a survival kit.

Army aircraft normally have survival kits on board for the type area(s) over which they will fly. There are kits for overwater survival, for hot climate survival, and for cold climate survival; there is also an aviator survival vest. As you go over the lists of equipment in these survival kits, you will note that each kit has–

- Water or a means to obtain/store water (desalter kit, water storage bag).
- Food packets and items you can use to obtain food (fishing tackle, snare wire).
- Trioxane fuel.
- Wooden matches and waterproof matchbox.
- Frying pan.
- Pocketknife.

- Plastic spoon.
- Compass.
- Signaling mirror (the overwater kit also contains sea dye marker, and the hot climate kit contains a plastic whistle).
- Mark 113 MOD O.
- First aid kit.
- Items to protect oneself against environmental elements (sunburn-preventive preparation,, insect headnet, and reversible sun hat are in the overwater and hot climate kits; poncho, sleeping bag, insect headnet, shovel, and candles are in the cold climate kit), and a survival manual.

If you are not an aviator, you will probably not have access to the survival vest or the survival kits, but becoming familiar with what they contain will help you to plan and to prepare your own survival kit.

Even the smallest survival kit if properly prepared is invaluable when you are faced with a survival problem.

Selection of Survival Kit Items

The key to the types of items you will need in your survival kit is the environment.

The key to how much you put in your kit and where you carry it is your mode of travel and your unit's organic equipment.

In preparing your survival kit, select items you can use for more than one thing. If you have two items that will serve the same function, pick the one that you can also use for another function. Do not duplicate items as this will increase the size and weight of your kit.

Your survival kit need not be elaborate. You need only functional items that will meet your needs and a case to hold the items. For the case you might use a band-aid box, a first aid case, an ammo pouch, or other suitable case, but it should be—

- Water repellent.

- Easy to carry.
- Durable, and
- Large enough to hold the items you need.

In your survival kit you should have—

- First aid items.
- Signaling items.
- Water and/or means to obtain potable water.
- Fire-starting items.
- Food and/or items to help you obtain food.
- Shelter items.

You should include a weapon only if the situation so dictates.

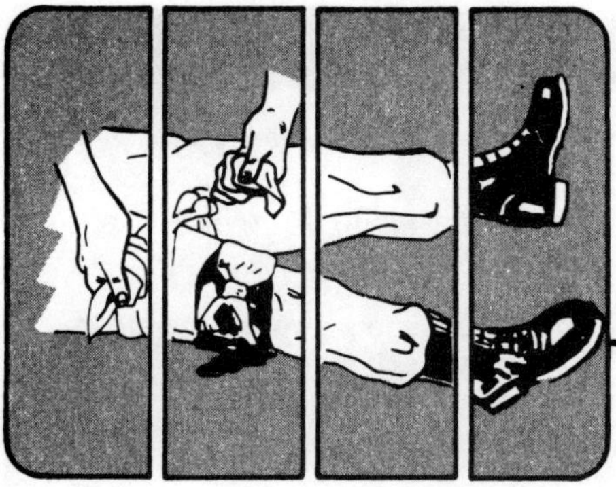

Chapter 3

Survival Medicine

You probably take good health for granted; at least, you take having a doctor available for granted. And when you feel sick, you head for the dispensary, knowing that the doctor will have a solution to your health problem: he can make you healthy.

In a survival situation, however, it is you who must know what to do to stay healthy. In other words, you must know how to avoid illnesses and how to prevent injuries.

Staying Healthy

Four of the most crucial factors in staying healthy are having adequate water and food, practicing good personal hygiene, getting sufficient rest, and having your shots/immunizations up to date.

Adequate Water. Your body loses water through normal body processes—sweating, urinating, and defecating. When the atmospheric temperature is 68° F, the average adult loses, and therefore requires, 2 to 3 liters (2 to 3 quarts) of water daily. In other circumstances, such as heat exposure, cold exposure, intense activity, high altitude, burns, or illness, your body may lose more water. This water must be replaced. Figure 3-1 describes what occurs when lost body fluid is not replaced.

Dehydration results from inadequate replacement of lost body fluids. It will decrease your efficiency in doing even the simplest task, and it will increase your susceptibility to severe shock if you are injured.

Thirst is no indication of how much water you need, so even when you are not thirsty, drink small amounts of water regularly to prevent dehydration. If you are exerting a lot of energy or are under

BODY FLUID LOSS	RESULTS
5%	Thirst, irritability, nausea, and weakness.
10%	Dizziness, headache, inability to walk, and a tingling sensation in limbs.
15%	Dim vision, painful urination, swollen tongue, deafness, and a numb feeling in the skin.
More than 15%	Death.

Figure 3-1. The effects of dehydration.

severe conditions, increase your water intake. Drink enough liquids to maintain a urine output of at least 1 pint every 24 hours. In a hot climate, you should drink 4 to 8 gallons of water a day.

To treat dehydration, replace the body fluids that were lost. Drink any potable fluids available—water, fruit juices, soft drinks, tea, and so forth. (Chapter 5 tells you ways to obtain water).

CAUTION: Do not use sea water or urine under any circumstances. Although they will satisfy thirst temporarily, they actually cause additional water loss from the body, promote dehydration, and if taken in sufficent quantity, will kill you.

Adequate Food. Although you can live several days without food, you need an adequate amount to stay healthy. Without food your mental and physical capability will rapidly deteriorate, and you will become weak. Food replenishes the substances that your body burns and provides energy. It provides vitamins, minerals, salts, and other elements essential to good health. Possibly more important, it helps morale.

Meats provide fats, which furnish energy, and proteins, which the body uses for production of special chemical compounds it needs. Plants provide carbohydrates, which are the main source of energy, and many plants provide enough protein to keep the body at normal efficiency.

In a survival situation, the type of food available may be unappealing and foul-smelling. But if it's edible, eat it. (See Chapters 6 and 7 for information on using wild plants and animals for food.)

Personal Hygiene. In any situation, cleanliness is an important factor in preventing infection and disease; in a survival situation it becomes even more important.

Of course, a daily shower with hot water and soap is ideal, but you can keep clean without this luxury. Use a cloth and soapy water to wash yourself. If water is scarce, take an "air" bath: Remove as much of your clothing as practical and expose your body to sun and air.

If you are out of soap, you can use ashes or sand. Or you can make soap from animal fat and wood ashes if your situation allows. The odor from the fire and hot fat can alert the enemy of your presence.

To make soap—

- Extract grease from animal fat by cutting the fat into small pieces and cooking them in a pot. Add enough water to the pot to keep the fat from sticking as it cooks. Cook the fat slowly, stirring frequently. After the fat is rendered, pour the grease into a container to harden.
- Place ashes in a container with a spout near the bottom. Pour water over the ashes, and in a separate container collect the liquid that drips out of the spout. This liquid is the potash or lye. Another method for obtaining the lye is to pour the slurry (the mixture of ashes and water) through a straining cloth.
- In a cooking pot, mix two parts grease to one part potash. Place this mixture over a fire and boil it until it thickens.
- After the mixture—the soap—cools, you can use it in the semiliquid state directly from the pot, or you can pour it into a pan, allow it to harden, and cut it into bars for later use.

Keep Your Hands Clean. Germs on your hands can infect food and wounds. So be sure to wash your hands after handling any material that is likely to carry germs, after visiting the latrine, after caring for the sick, and before handling any food, food utensils, or drinking water. Keep your fingernails closely trimmed and clean, and keep your fingers out of your mouth.

Keep Your Hair Clean. Your hair can become a haven for fleas, lice, and other parasites or bacteria. Keeping your hair clean and trimmed will help you to avoid this danger. Fleas and lice live and feed on warm-blooded animals and are carriers of dangerous diseases. Rodents, for instance, are likely to have fleas or lice. So if you kill a rodent to eat, let it become completely cold before cleaning it so that the lice or fleas will be gone. Louse powder is the best way to rid yourself of fleas and lice should you become infested. Other ways to rid yourself of these insects is to place your clothing in direct sunlight for a few hours or wash frequently in hot, soapy water.

Keep Your Clothing Clean. You should keep your clothing and bedding as clean as possible to reduce the chance of skin infection as well as to decrease the danger of parasite infestation. Clean your outer clothing whenever it becomes soiled. Wear clean underclothing and socks each day. If water is in short supply, "air" clean your clothing (shake, air, and sun for 2 hours). Turn your sleeping bag inside out after each use and fluff and air it.

Keep Your Teeth Clean. At least once each day, thoroughly clean your mouth and teeth with a toothbrush or dentifrice. If you don't have a toothbrush, make a "chewing stick." Find a twig about 8 inches long and ½ inch wide. Brush your teeth thoroughly with your chewing stick. Or wrap a clean strip of cloth around your finger and rub your teeth with it to wipe away food particles. To remove food stuck between your teeth, use a toothpick made from a twig or use dental floss made from thread, string, or thin strips of bark or vine.

Take Care of Your Feet. To prevent serious foot trouble, break in your shoes prior to wearing them on any mission, wash and massage your feet daily, trim your toenails straight across, and check your feet for blisters. If you get a blister, do not open it. An intact blister is safe from infection. Apply a dressing around the blister, not on it. If the blister bursts, clean it and apply a bandage over it.

Sufficient Rest. You need a certain amount of rest to keep going. Rest restores physical and mental vigor. If you are ill or injured, it promotes healing. You should plan for regular rest periods in your daily activities, and you should learn to make yourself comfortable under less than ideal conditions.

Immunizations. Before you go on a mission, make sure your immunizations are up to date.

Knowing how to treat and immediately treating a companion who has a serious medical problem are important in any situation; they are critical in a survival situation where medical personnel are not available and may not be available for weeks, possibly months. Stopped breathing, severe bleeding, and shock constitute serious medical emergencies that must be treated at once to prevent loss of life.

Stopped Breathing. Any one of the following can cause airway obstruction, resulting in stopped breathing:

- The presence of some foreign matter in the mouth or throat that obstructs the opening to the trachea.
- Face and/or neck injuries.
- Inflammation and swelling of the mouth, throat, and trachea caused by inhaling of smoke, flames, or irritating vapors or by an allergic response to foodstuffs, insect bites, plants, or other items.
- Flexion of the neck. If the neck is bent forward so that the chin is resting upon the chest, the resulting "kink" in the throat can block the passage of air from the mouth and/or nose to the lungs.
- Unconsciousness. A state of unconsciousness produces complete relaxation of the muscles of the lower jaw and the tongue. If the neck is bent forward, the lower jaw sags and the tongue drops back, blocking the passage of air into the lungs.

The following are signs of airway obstruction:

- The victim is struggling to move air in and out of his lungs.
- The muscles in the front of the victim's neck stand out prominently but no air can be heard or felt moving in and out of the mouth or nose.
- Cyanosis, a notable blue or gray color of the skin around the lips, ears, fingernails, and sometimes the whole body, is present.

Stopped breathing, regardless of its cause, is a critical medical emergency. When the air supply to the lungs is restricted or stopped, irreversible brain damage and eventually death can occur within a matter of minutes. *Time is of the utmost importance.*

Opening the Airway. To restore and maintain an open airway, take the following steps.

NOTE: *If any step opens the airway so that the victim begins to breath spontaneously, you need not proceed any further.*

Step 1. Clean out the victim's mouth. Using a finger, quickly sweep the victim's mouth clear of any foreign objects, broken teeth, dentures, sand, etc.

Step 2. Tilt the head back.

Adjust the victim's position to enlarge the airway by immediately placing him on his back with his head in a chin-up position (figure 3-2).

Tilt the victim's head backwards as far as possible so that the front of the neck is stretched tightly. This is done by placing one hand on the back of the neck and lifting while pushing down on the forehead with the heel of the other hand. This usually opens the mouth automatically.

If a rolled blanket, poncho, or similar object is available, it can be placed under the victim's shoulders to maintain this position, but do not waste time obtaining such materials.

If tilting the head back opens the airway and the victim starts to breathe, go no further; otherwise go on to step 3. *Keep the head tilted back throughout all of the remaining steps.*

Step 3. Force air into the lungs.

Try to force two or three breaths quickly into the victim's lungs through the mouth while holding the nostrils pinched shut. Watch the victim's chest for movement indicating that the air is reaching his lungs.

This forced breathing may be enough to start spontaneous respiration, or it may remove a small obstruction that has been restricting breathing.

Watch the victim's chest for movement indicating that the air is reaching the lungs. If the chest rises and falls with the forced ventilations, the airway is unobstructed. If the airway is not open, go on to Step 4.

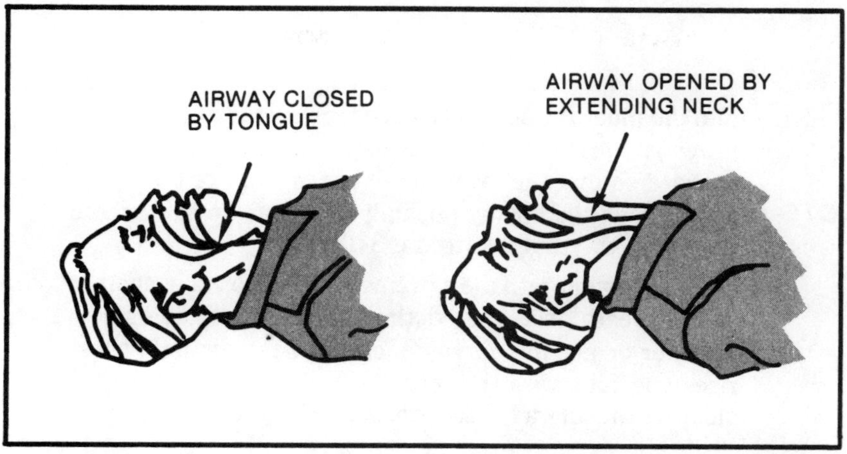

Figure 3-2. Neck extended in a chin-up position.

Step 4. Lift the jaw.

Accentuate the stretch of the neck to get the tongue out of the way by using one of the following two jaw-lift methods:

- Thumb jaw-lift. Place your thumb in the victim's mouth, grasp the lower jaw firmly with your fingers, and lift the jaw forward (figure 3-3). Do not try to hold or depress the tongue.
- Two-hand jaw-lift. This method is used when the victim's jaws are closed so tightly that the thumb cannot be inserted

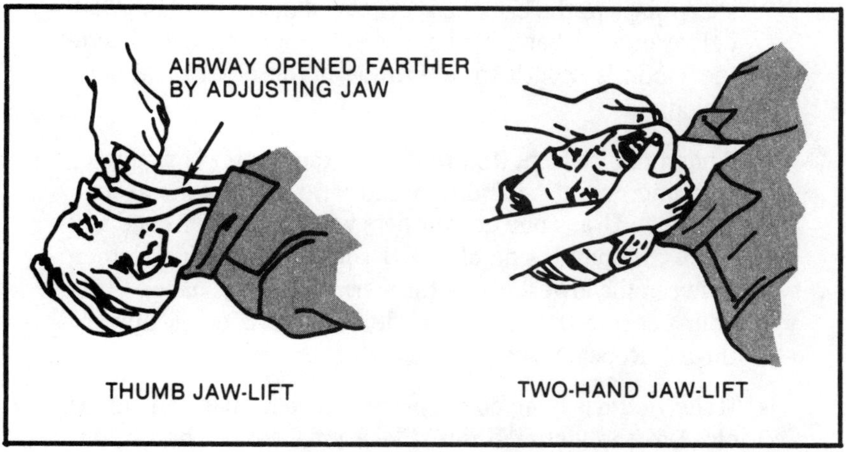

Figure 3-3. Method for adjusting lower jaw to jutting out position.

into the mouth. Using both hands, grasp the angles of the lower jaw just below the ear lobes. Lift the jaw forcibly forward so that the lower teeth are in front of the upper teeth. Open the lips by pushing the lower lip toward the chin with the thumbs (figure 3-3).

- Once the tongue is in the extreme forward position, another quick breath into the victim's mouth will determine whether or not the airway is clear. If the victim's chest rises and falls with the forced ventilation, the airway is clear. If the airway if still obstructed, proceed to step 5.

Step 5. Clear the air passage. When efforts to open the airway by head tilt, forced ventilation, and maximum jaw extension all fail, it is probably that a foreign object is lodged so deeply in the victim's throat that the quick sweep of the mouth in step 1 failed to reach it. The following maneuvers should be performed in an attempt to dislodge the object.

- Finger probe. Sweep an index finger down the inside of the victim's upper cheek to the base of the tongue. Use the finger as a hook to attempt to dislodge the foreign body up into the mouth. When the object comes within reach, remove it.

- Back blows. Turn the victim on his side and deliver a few sharp slaps to the back between the shoulder blades. After delivering the back blows, sweep an index finger inside the victim's mouth to determine if the object has been dislodged.

- Abdominal thrusts (figure 3-4). If the victim is sitting or standing, stand behind him and wrap your arms around his waist. Grasp one of your fists with your other hand and place the thumb side of this fist in the victim's abdomen between the lowest end of the sternum and the navel. Press your fist into the victim's abdomen with a quick upward thrust. Repeat several times if necessary.

- If the victim is lying down, place him on his back and kneel close to the victim's hips. Place the heel of one of your hands against the victim's abdomen, between the lowest end of the sternum and the navel. Place your other hand on top of the first. Press your hand into the victim's ab-

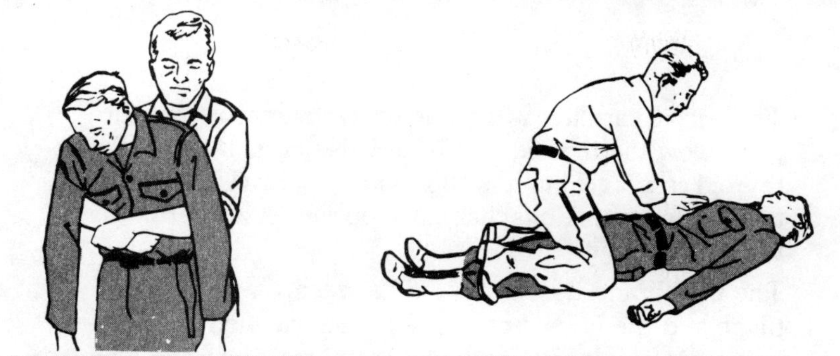

Figure 3-4. Abdominal thrust.

domen with a quick upward thrust. Repeat several times if necessary.

- Combined use of back blows and manual thrusts. A combination of back blows and manual thrusts may be necessary for clearing upper airway obstructions.

Cricothyroidotomy. A person with acute upper airway obstructions may require an immediate cricothyroidotomy (an artificial airway) to save his life. A cricothyroidotomy is an opening in the trachea between the thyroid cartilage and the cricoid cartilage that permits air to pass directly from the outside into the trachea without passing through the upper air passages. A cricothyroidotomy is performed as follows:

CAUTION: This procedure requires specific knowledge and training. Nonmedical personnel should perform it only in a combat survival situation and only as a last resort.

- Place the victim on a flat surface with the neck extended and the head tilted backwards so that the structures in the neck are stretched.
- Only if time permits, clean the skin with soap and water. If an antiseptic is available, apply it to the skin.
- With your fingers on the victim's neck, identify precisely the cricothyroid membrane (figure 3-5).

  The thyroid cartilage is the rather large structure that forms the Adam's apple.

The cricoid cartilage is the ringlike cartilaginous structure just below the thyroid cartilage. It is not as large as the thyroid cartilage, but it is approximately twice as thick as the cartilaginous rings that make up the remainder of the trachea below it.

The membrane that is located between the thyroid cartilage and the cricoid cartilage is the cricothyroid membrane. At this point, only the membrane and the skin separate the airway from the outside.

- Lift the skin over the cricothyroid membrane and make a 1/2-inch vertical incision through the skin in the midline with a scalpel, razor blade, knife, or any other sharp instrument.

- With your fingers, separate the skin, exposing the cricothyroid membrane, and make a transverse incision in the membrane, exposing the tracheal lumen.

  Once the incision has been made in the cricothyroid membrane, the opening *must* be maintained so that air can move between the trachea and the outside. This can best be done by carefully inserting a piece of clean tubing, which can be obtained from a ballpoint pen, into the opening; however, in an absolute emergency, *anything* that will maintain this opening may be used. When the tube has been inserted, air can be heard moving in and out of the opening immediately.

- Secure the opening. If a tube has been inserted, attach a strip of cloth or a piece of string carefully around the tube and tie it around the neck. The tube must be anchored to keep it from falling out or from being jammed into the back wall of the trachea.

  The opening must be maintained until the victim is in the care of a physician or until the unconscious victim recovers enough that the cricothyroidotomy is no longer necessary (swelling goes down and victim can breathe normally). When the tube is removed, the wound will close and heal with no further attention, other than antiseptic and an airtight bandage.

Bleeding. Severe bleeding from any major blood vessel in the body is extremely dangerous. Loss of 1 quart of blood will produce moderate symptoms of shock, loss of 2 quarts will produce a severe state of shock that places the body in extreme danger, and loss of 3 quarts is usually fatal. In a survival situation, serious bleeding must be controlled *immediately* because replacement fluids normally are not available and the victim can die within a matter of minutes.

External bleeding can be classified according to its source.

Arterial. Blood is carried away from the heart and through the body in blood vessels called arteries. When an artery has been cut, *bright red* blood issues from the wound in *distinct spurts* or pulses that correspond to the rhythm of the heartbeat. Because the blood in the arteries is under high pressure, quite a large volume of blood can be lost in a short period of time when an artery of significant size is damaged. For this reason, arterial bleeding is considered the most serious type of bleeding; if it is not controlled promptly, it can be fatal.

Venous. Venous blood is blood that is being returned to the heart by means of blood vessels called veins. Bleeding from a vein is characterized by a *steady flow of dark red, maroon,* or *bluish* blood. Pressure of the blood in the veins is much lower than in the arteries, but nevertheless, venous bleeding can be quite profuse. Venous bleeding usually is controlled more easily than arterial bleeding.

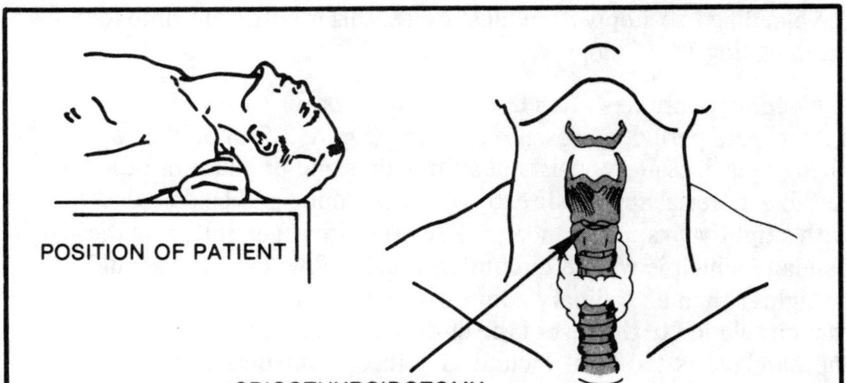

Figure 3-5. Location of cricothyroid membrane.

Capillary. The capillaries are the extremely small vessels that connect the arteries with the veins. Capillary bleeding most commonly occurs in minor cuts and scrapes. Blood flow from such injuries is *slow* and *oozing,* and blood flow is not profuse. Bleeding from capillaries is not difficult to control because the blood usually clots and the bleeding stops on its own accord.

Control external bleeding by direct pressure, elevation, or tourniquet.

Direct pressure. The most effective means of controlling external bleeding is by the application of pressure directly over the wound. This pressure must not only be firm enough to stop the bleeding, but it also must be maintained long enough to "seal off" the damaged surface.

The initial application of pressure should be with a finger or hand placed directly over the bleeding point. If a sterile dressing or clean cloth compress is immediately available, it may be held over the bleeding point as the pressure is applied, but time should not be wasted locating these.

Firm, even pressure should be applied to the bleeding point until the bleeding stops. However, alternate application and relaxation of the pressure to determine if the bleeding has stopped is not desirable. It is best to apply the pressure to the bleeding point continuously for up to 30 minutes before releasing it to determine if the bleeding has stopped. In most cases, this is sufficient time for the bleeding to be stopped.

If bleeding continues after the application of direct pressure for a 30-minute period, a pressure dressing (figure 3-6) should be applied. This dressing consists of a thick dressing of gauze or other suitable material applied directly over the wound and held in place with a tightly wrapped bandage. The pressure is generated by the bandage which is wound circumferentially. This bandage should be tighter than an ordinary compression bandage but not so tight that circulation to the rest of the limb is impaired. Indications that the bandage is too tight include the absence of pulses, a bluish discoloration of the nails and skin, and tingling sensations or pain in the extremity beyond the bandage.

Once a pressure dressing has been applied it *should not be removed* even if bleeding continues. If the dressing becomes blood-soaked, sufficient pressure was not generated to stop the bleeding and additional pressure must be applied. This additional pressure may be provided by another dressing placed on top of the original dressing. Elevation of the wounded extremity and application of digital pressure should be used in conjunction with the additional dressing.

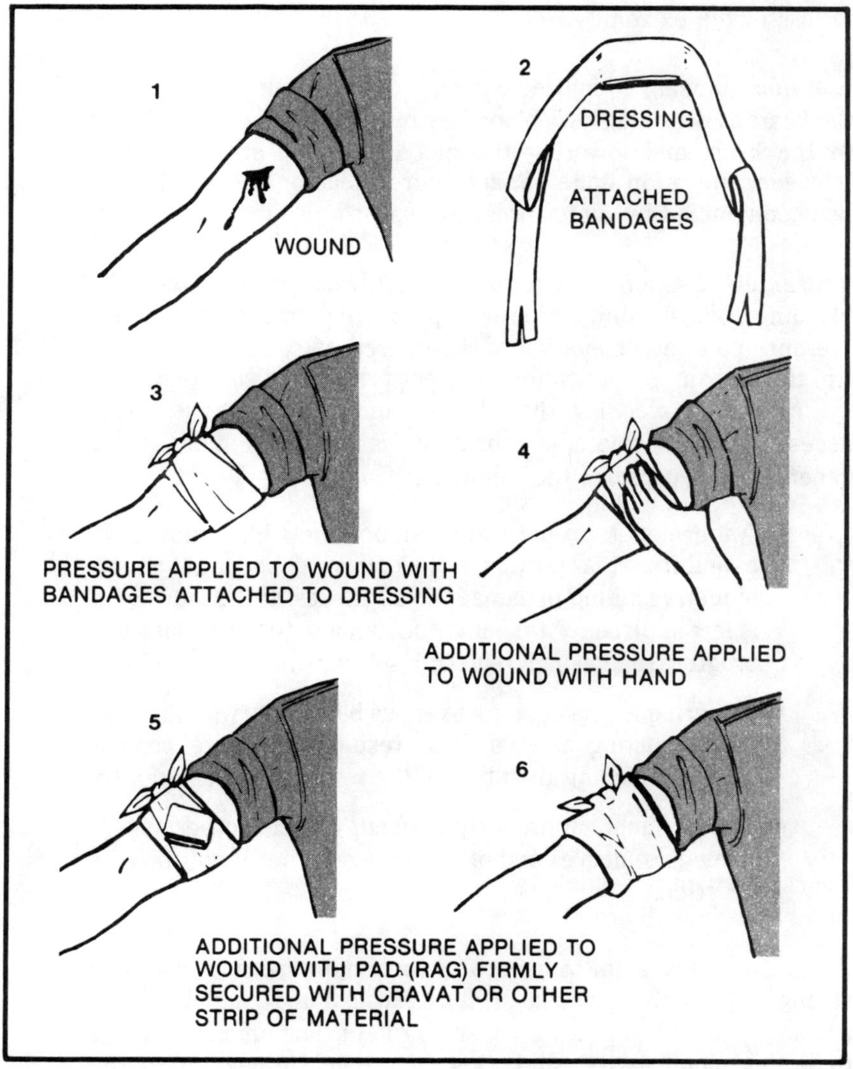

Figure 3-6. Application of a pressure dressing.

The pressure dressing should be left in place for 1 or 2 days, after which it can be removed and replaced by a smaller dressing. During the 2-day interval, the wound and pressure dressing should be checked frequently for indications of controlled bleeding, insufficient circulation, and infection. Failure to do this could result in gangrene or frostbite, due to a decrease in circulation, and then in loss of the extremity.

Elevation. Raising an injured extremity as high above the level of the heart as possible slows blood loss by aiding the return of blood to the heart and lowering the blood pressure at the wound. However, elevation alone *will not* control bleeding entirely; it *must* be accompanied by direct pressure over the wound.

Tourniquet. Use a tourniquet only when direct pressure over the bleeding point in conjunction with pressure over the appropriate pressure point and elevation of the injured extremity fail to control the bleeding. Application of direct pressure is so effective in the control of bleeding that the use of a tourniquet is rarely necessary; furthermore, a tourniquet is not recommended for general use because of the following:

- A tourniquet, properly applied, obstructs blood flow both to and from that portion of the extremity beyond the tourniquet, resulting in damage to all tissues. If the tourniquet is left in place for too long, the damage to the tissues can progress to total gangrene with subsequent loss of the limb.
- A tourniquet may obstruct venous blood flow without totally obstructing arterial flow, resulting in more profuse arterial bleeding than before the tourniquet was applied.
- A tourniquet, improperly applied, can cause permanent damage to nerves and other tissues at the site of the constriction.

If you must use a tourniquet, you can improvise one from any strong, soft, pliable material such as gauze, a large handkerchief, a triangular bandage, a towel, or similar item. To minimize damage to nerves, blood vessels, and other underlying tissues, the tourniquet should be 3 to 4 inches wide before it is wrapped around the extremity and at least 1 inch wide after it is tightened. Apply the tourniquet as follows:

- Place the tourniquet around the extremity between the wound and the heart 2 to 4 inches above the wound site (figure 3-7). Never place it directly over the wound or directly over a fracture.
- Wrap the tourniquet twice around the extremity and tie it with a single overhand knot. Place short stick or similar object on top of this knot (figure 3-7) and use a square knot to hold it in place.
- Use the stick as a handle to tighten the tourniquet (figure 3-7). Tighten it only enough to stop blood flow. If you can feel a pulse in the intact extremity before the tourniquet is applied, use the absence of this pulse after tightening the tourniquet as an indicator that tourniquet pressure is sufficient.
- When you have tightened the tourniquet sufficiently, bind the free end of the stick to the limb to prevent unwinding (figure 3-7).
- After you secure the stick, clean, dress, and bandage the wound.

A lone survivor should not remove or release a tourniquet once it has been applied.

Shock. Shock is not a disease entity in itself; it is a syndrome, or collection of symptoms. These symptoms result from a decrease in the volume of blood circulating through the body and from the body's efforts to compensate for this decrease.

Signs/symptoms. Early signs of shock are—

- Pale skin.
- Rapid pulse.
- Coldness of extremities.
- Thirst.
- Dryness of mouth.

Symptoms of shock often observed are—

- Faintness, disorientation, or unexplainable restlessness.
- Weak, feeble, pulse.

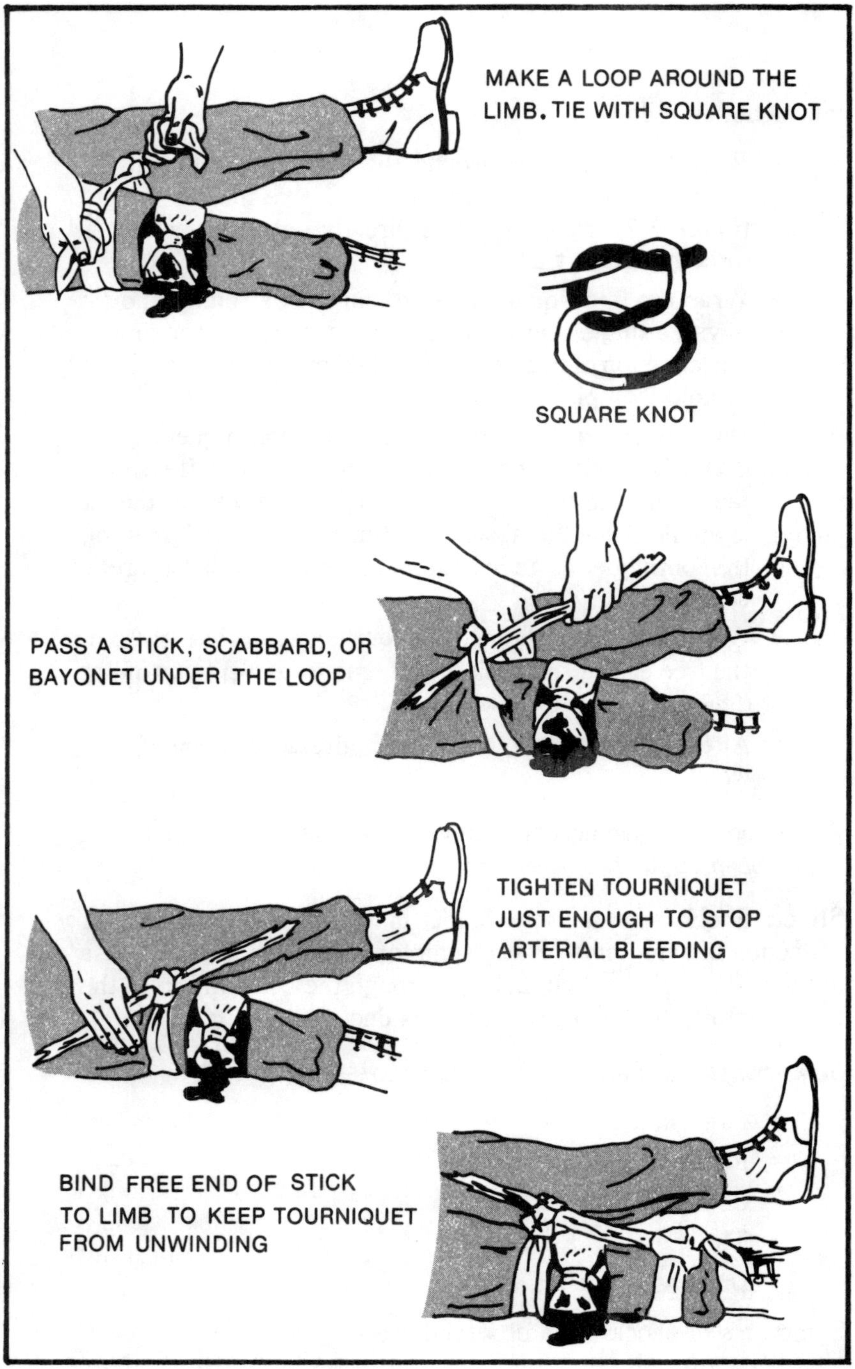

Figure 3-7. Application of tourniquet.

- Cold, clammy sweating.
- Decreased urinary output.

As shock becomes more severe, the following symptoms will be evident:

- Rapid and weak or "thready" pulse, or unobtainable pulse.
- Irregular, gasping respirations.
- Dilated pupils that are slow to respond to light.
- Mental confusion and eventual coma and death.

Complications. Shock can cause death if it is not treated promptly, even though the injuries that brought on shock might not be serious enough to cause death. In fact, individuals may go into shock without having any physical injuries.

Treatment. Shock should be anticipated in all persons who have been injured. All injured persons should be treated in the following manner, regardless of what symptoms appear, in order to prevent and/or control shock:

- If the victim is conscious, place him on a level surface with the lower extremities elevated 15 to 20 cm (6 to 8 inches).
- If the victim is unconscious, place him on his side or abdomen with the head turned to one side to prevent choking on vomitus, blood, or other fluids.
- If there is any doubt as to the best position, place the victim perfectly flat. Once the victim is in a shock position, do not move him.
- Maintain body heat by insulating the victim from the surroundings and, in some instances, applying external heat.
- If the victim is wet, remove all wet clothing as soon as possible and replace with dry garments.
- Insulate the victim from the ground with clothing, parachute material, tree boughs, or any other available material. Improvise a shelter to insulate the victim from the weather.
- Use warm liquids or foods, a prewarmed sleeping bag, another person, warmed water in canteens, hot rocks wrapped in clothing, or fires on either side of the victim to provide external warmth.

The warm liquids or foods should be given only if the victim is conscious.

- If the victim is conscious, administer a warm salt solution slowly in small doses to him. The use of bouillon, tea, or

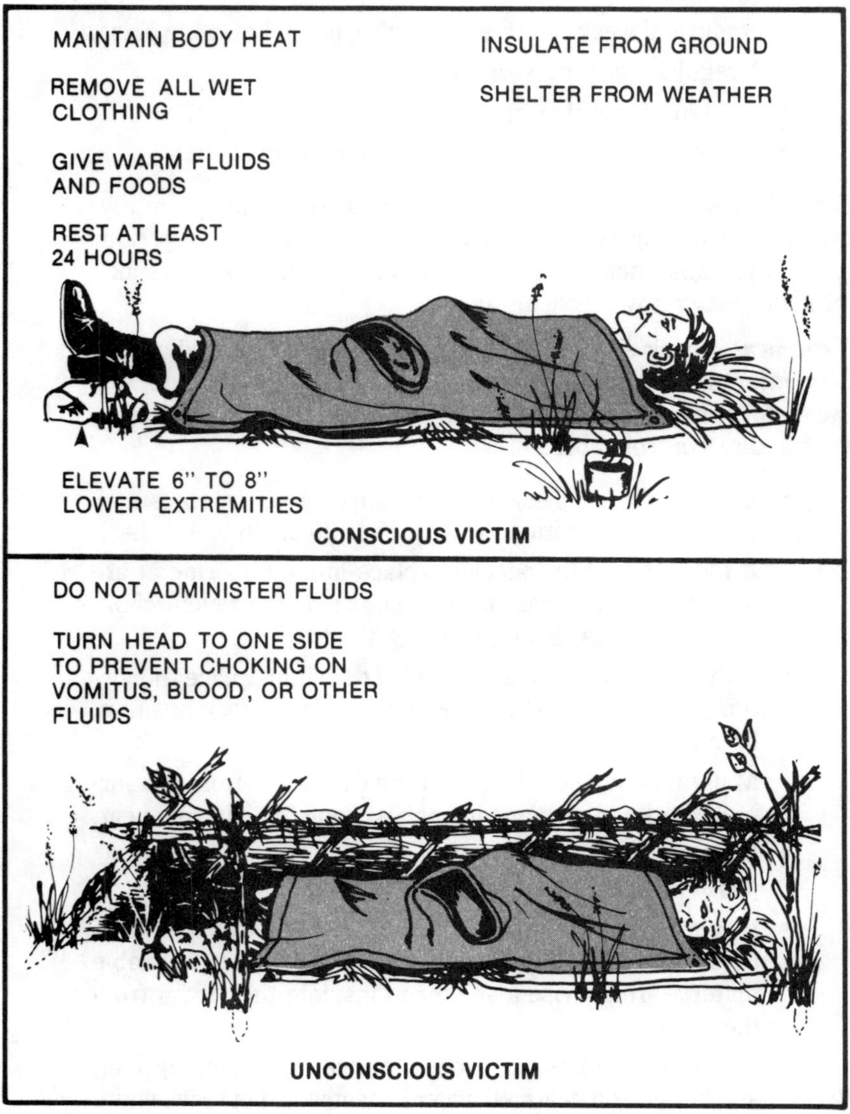

Figure 3-8. Treatment for shock.

any other warm beverage may make the solution more palatable.

- Do not administer fluids if the victim is unconscious or if the victim has abdominal wounds.
- The victim should rest for at least 24 hours.

If you are a lone survivor, lie in a depression in the ground, behind a tree, or any other place out of the weather with head lower than feet. Try to keep warm and to rest for at least 24 hours (see figure 3-8).

Other Health Hazards

Insect bites and stings, animal bites, open wounds, skin infections, and intestinal parasites are just a few of the health hazards that you may face in a survival situation. Learn how you can avoid these hazards (Chapters 13-16) and how to treat yourself if you become a victim to these health problems.

Insect Bites and Stings. Insects and related pests are hazards in a survival situation not only because of the irritation that results but because some are carriers of diseases and others may cause a severe allergic reaction in some individuals. In many parts of the world you will be exposed to serious, even fatal, diseases not encountered in the United States. For instance—

- Mosquitoes may carry malaria, dengue, and many other diseases.
- Flies can spread disease from contact with infectious sources. They are causes of sleeping sickness, typhoid, cholera, and, in rare cases, dysentery.
- Fleas can transmit plague.
- Lice can transmit typhus and relapsing fever.
- Ticks can carry and transmit diseases, such as Rocky Mountain spotted fever which is common in many parts of the United States.
- Bee and wasp stings can be dangerous and even fatal in individuals who are sensitive to their venom.

If you get bitten or stung, do not scratch the bite as it may become infected.

You should inspect your body at least once a day to ensure that there are no insects attached on you. If you find ticks attached to your body, cover them with a substance such as vaseline, heavy oil, or tree sap that will cut off their air supply. This will cause the tick to release its hold and you can remove it. Take care to remove the whole tick. Use tweezers if you have them. Grasp the tick where the mouth parts are attached to the skin. Do not squeeze the tick's body. Wash your hands after touching the tick since infective fluids from it can cause disease. Clean the tick wound well each day until it is healed.

If you have been in a chigger or mite-infested area, wash your skin thoroughly with soap and water, repeating several times.

If you are stung by a bee or wasp, immediately remove the stinger and venom sac if attached by scraping with a fingernail or a knife blade. Do not squeeze or grasp the stinger or venom sac with tweezers or with your fingers as squeezing will force more venom into the wound. Wash the sting site thoroughly with soap and water to lessen the chance of a secondary infection and apply an ice pack or cold compress.

If you know or suspect you are allergic to insect stings, always carry an insect sting kit with you.

You can sometimes relieve the itching and discomfort caused by insect bites by applying cold compresses, a cooling paste of mud and ashes, the milky sap from dandelions, coconut meat, or crushed leaves of garlic.

Other pests whose bites or stings can be dangerous are spiders, centipedes, scorpions, and ants. A few spiders have poisonous bites that may be as painful as a wasp's sting. A centipede's sting is like that of a wasp. A scorpion's sting can make you sick or kill you, depending on the species. Biting ants can cause a lot of discomfort and serious allergic reactions in some people. To treat a spider or scorpion bite, clean the wound extremely well and try to remove the toxin by suction or by squeezing the bite site. If you have any type of tobacco, chew it and place it over the bite site. This will help to ease the pain. Treat the bite as you would an open wound (page 3-24).

Animal Bites. Although you are less likely to be bitten by an animal than an insect in a survival situation, it is extremely important that you know how to treat an animal bite. If there are no means to test the animal for rabies, consider its bite rabid and begin shots if serum is available. Clean the wound extremely well, and then continue to treat the bite as you would an open wound (page 3-24).

NOTE: *Consider using the treatments the local population uses on spider and animal bites, but do so with caution (page 3-20).*

Snakebite. The possibility of a snakebite in a survival situation is rather small if you are familiar with the various types of snakes and their habitats. But because the possibility does exist, you should know how to treat a snakebite. You should also know that deaths from snakebites are rare. More than one-half of the victims of snakebite will have little or no poisoning, and only about one-quarter will develop serious systemic poisoning. However, the possibility of snakebite in a survival situation can affect morale, and failure to take preventive measures or failure to properly treat a snakebite can result in needless tragedy.

A bite wound, regardless of the type of animal that inflicted it, can become infected from the bacteria in the animal's mouth. With non-poisonous as well as poisonous snakebites, this local infection is responsible for a large part of the residual damage that results.

Shock and panic in a person bitten by a snake can also affect the person's recovery. Excited, hysterical, and panicky reactions can speed up the circulatory system, causing the body to absorb the toxin quickly.

Very few snakebite victims die or become permanently disabled, however, if they receive proper treatment promptly.

Before you begin treatment for a snakebite, you should determine if the bite was made by a poisonous or nonpoisonous snake. If you did not get a good enough look at the snake to identify it , you can tell by the bite mark whether it was poisonous.

All snakes have several rows of teeth. Some snakes, however, have one or more pairs of fangs (large, grooved or hollow teeth) through which the snake injects venom. All snakes that possess fangs are considered poisonous.

Look for the following signs/symptoms to determine if a snakebite is made by a poisonous snake:

- Fang punctures at the site of the bite. One or two punctures (sometimes three or four) generally indicate a poisonous snake, especially if one or more of the symptoms below are present.

- Pain at the site of the bite.
- Swelling at the site of the bite within a few minutes or within 2 hours.
- Paralysis, weakness, twitching, and numbness. These are signs of neurotoxic venoms, and usually appear 1½ to 2 hours after the person is bitten.

If you determine that you have been bitten by a poisonous snake, take the steps listed in figure 3-9.

If antivenom is available and *you are certain you are not allergic to it, use it.*

If a medically trained person (medic or doctor) is present, he should give the bite victim a 0.1 cc test dose of antivenom. If the victim is allergic to the antivenom but IV fluids are available, the medic should start two IV lines, TKO rate, one with antivenom, then one with epinephrine of 1:1000 0.4-1 cc. (Use epinephrine only if allergic reaction occurs.) Repeat every 5 to 10 minutes if needed. Keep IV dripping.

If the victim is allergic to the antivenom and no IV fluid is available, divide the dose of antivenom.

CAUTION: In a survival situation, the amount of antivenom available will be minimal.

If the victim is allergic to the antivenom, using one or two vials will probably cause more danger than help due to anaphylactic reaction. To be effective for a major envenomation, 15 to 20 vials of antivenom are required.

Following the above-described care of the victim, the medic should take the following actions to minimize local effects:

- Remove any necrotic tissue.
- Watch for infection.
- If infection appears, keep the wound open and clean. Remove necrotic tissue as needed. If scrubbing does not remove all necrotic tissue and the infection is getting worse, consider maggot therapy (page 3-25).
- May use sterile (or as clean as possible) cloth as a wick for continued drainage.

- Flush the wound daily with water or fresh urine. (Do not store urine for later use.)
- Use heat after 24 to 48 hours to help prevent spread of local infection.
- Keep wound covered with dry sterile dressing.
- Have victim drink large amounts of fluids until infection is gone.

DO'S	DON'TS
Lie down (if circumstances permit) with head slightly lower than the rest of the body.	Do not move around because activity makes the blood circulate faster thereby speeding up the spread of the venom.
Refer to the DON'TS before considering cutting. If making an incision, cut no deeper than the two layers of skin.	Do not make any deep cuts at bite site. Cutting opens capillaries, which in turn opens a direct route into the blood stream for venom and infection.
Remove toxin as soon as possible by using a mechanical suction device or by squeezing.	Do not use your mouth to suck out venom. The vessels under the tongue will absorb toxins almost immediately and carry them to the heart.
Clean the bite site and hands extremely well.	Do not use ice on bite.
If the bite is on an extremity, snuggly wrap the extremity halfway above the bite site, using material at least 2 inches wide.	Do not put hands on face as venom may be on hands.
Remove watches, rings, bracelets, and any other restricting items.	Do not use a tourniquet.
Drink small amounts of fluid, preferably water.	Do not drink alcoholic type fluids.
Use aspirin or Tylenol for pain.	Do not use morphine or morphine derivatives.

Figure 3-9. Snakebite do's and don'ts.

Open wounds. All open wounds are serious in a survival situation, not only because of tissue damage and blood loss, but also because they may become infected. Bacteria on the object that made the wound, on the skin and clothing of the individual, or on other foreign material or dirt that touches the wound may cause infection.

By taking proper care of the wound, however, you can reduce further contamination and promote healing. Clean the wound as soon as possible after it occurs:

- Remove or cut clothing away from the wound.
- Thoroughly clean the skin *around the wound.*
- Rinse (do not scrub) the wound with large amounts of the cleanest water available. You can use *fresh* urine if water is in short supply. Fresh urine is sterile.

CAUTION: *Do not apply common antiseptics such as iodine, merthiolate, and mercurochrome directly to a wound without diluting them. These solutions can cause further tissue damage, and should be used on intact skin only.*

CAUTION: *Do not attempt to close a wound by sewing or by a similar procedure. Leave the wound open to allow pus or infectious material to drain.*

- Cover the wound with a clean dressing. Place a bandage on the dressing to hold it in place.
- Change the dressing daily to check for infection.

In a survival situation, some degree of wound infection is almost inevitable. Pain, swelling, and redness around the wound; increased temperature; and pus in the wound or on the dressing indicate infection is present.

To treat an infected wound—

- Place a warm, moist compress directly on the infected wound. Change the compress when it cools, keeping a warm compress on the wound for a total of 30 minutes. Apply the compresses three to four times daily.
- Drain the wound. Open and gently probe the infected wound with a sterile instrument (knife, glass, wood, or other item) so that the pus can drain. Gently remove all accumulations of pus or crusted matter.

- Dress and bandage the wound.
- Drink a lot of water.
- Continue the above treatment daily until all signs of infection have disappeared.

During World War I, maggots were an accepted treatment for infected wounds. However, you should be aware of the following before you decide to use maggots to eat infected tissue:

- You must expose the wound to flies to introduce maggots. Because of their filthy habits, flies are likely to introduce other bacteria to the wound, possibly causing more complications.
- Maggots will invade live, healthy tissue when the dead tissue is gone or not readily available.

You should consider maggot therapy despite its hazards when you do not have antibiotics and the wound has become severely infected, does not heal, and ordinary debridement is impossible.

To use maggot therapy, proceed as follows:

- Expose wound to flies for one day and then cover wound.
- Check daily for maggots.
- Once maggots develop, keep wound covered but *check* daily.
- Remove all maggots once they have cleaned out all dead tissue and *before* they start on healthy tissue. Increased pain and bright red blood in the wound indicate that the maggots have reached healthy tissue.
- Flush the wound repeatedly with sterile water or fresh urine to remove the maggots.
- Check the wound every 4 hours for several days to ensure all maggots have been removed.
- Bandage the wound and treat it as any other wound. It should heal normally.

Skin Infections. Although boils, fungal infections, and heat rash rarely develop into a serious health problem, they cause discomfort, and you should treat them.

Boils. Apply warm compresses to bring the boil to a head. Then open the boil using a sterile knife, wire, needle, or similar item.

Thoroughly clean out the pus using soap and water. Cover the boil site, checking it periodically to ensure no further infection develops.

Fungal infections. Keep the skin clean and dry, and expose the infected area to as much sunlight as possible. *Do not scratch* and do not use strong substances such as iodine and alcohol. You cannot "burn out" fungus.

Heat rash. Keep the area clean and dry. Apply powder if available. A cold compress may help relieve itching.

Intestinal Parasites. You can usually avoid infestations of worms and other intestinal parasites if you take preventive measures. For example, never go barefooted. The most effective way to avoid intestinal parasites is: Do not eat uncooked meat and do not eat raw vegetables possibly contaminated with raw sewage or human feces used as a fertilizer (night soil). However, if you should become infested and lack proper medicine, you might try one of the following home remedies. Keep in mind, though, that these home remedies are not without danger. Most of these following remedies work on the principle of changing the environment of the gastrointestinal tract:

- Saltwater. Mix 4 tablespoons of salt in 1 quart of water and drink. *Do not* repeat this treatment.
- Tobacco. Eat 1 to 1½ cigarettes. The nicotine in the cigarette will kill or stun the worms long enough for your system to pass them. If the infestation is severe, repeat the treatment in 24 to 48 hours, *but not sooner.*
- Kerosene. Drink 2 tablespoons of kerosene but no more. If necessary, you can repeat this treatment in 24 to 48 hours but no sooner.
- Hot peppers. Peppers are effective only if they are a steady part of your diet. You can eat them raw or put them in soups or rice and meat dishes.

Diarrhea. This is a common, debilitating ailment that may be caused by such things as a change of water and food, drinking contaminated water, eating spoiled food, becoming fatigued, and using dirty dishes. You can avoid most of these causes by practicing preventive medicine. If you get diarrhea, however, and do not have antidiarrhea medicine with you, you may find one of the following treatments is effective:

- Limit your intake to fluids for 24 hours.
- Drink 1 cup of a strong solution of tea every 2 hours until the diarrhea slows or stops. The tannic acid in the tea helps to control diarrhea. Tannic acid is also found in the moist inner bark of hardwood trees. Boil the inner bark for 2 hours or more to release the tannic acid. Although this solution will have a vile taste and smell, it will stop most cases of diarrhea.
- Make a solution of one handful of ground chalk, charcoal, or dried bones and treated water. (The solution should be the consistency of kaopectate.) If you have some apple pomace or the rinds of citrus fruit, add an equal portion to the mixture to make it more effective. Take 2 tablespoons of the solution every 2 hours until the diarrhea slows down or stops.

You should drink a lot of liquids to replace the lost fluids.

Burns. Following is a field treatment for burns that relieves the pain somewhat, seems to help speed healing, and offers some protection against infection:

- Soak dressings or clean rags for 10 minutes in a boiling tannic acid solution (tea or inner bark of hardwood trees boiled in water).
- Cool the dressings or clean rags and apply over the burns.

CAUTION: Do not apply grease or fats to burns.

Herbal Medicines

Our modern day wonder drugs and fine laboratories and equipment have obscured the old-time "country doctor" type of medicine—determination, common sense, and a few primitive treatments. In many areas of the world, however, the people still depend on the local witch doctor or healer to cure their ailments. And many of the herbs (plants) and treatments they use are as effective as the most modern medications available. In fact, many modern medications come from refined herbs. For example, the white, stringy part of the rinds of citrus fruits and the apple pulp (apple pomace) left after pressing out the juice contain pectin. By mixing either the rind pieces or the apple pomace with ground chalk, you will have a primitive form of kaopectate.

You should use herbal medicines with extreme caution, however, and *only when you lack* or have limited *medical supplies.* Some herbal medicines are dangerous and may cause further damage or even death.

Chapter 4

Field Expedient Weapons and Tools

As a soldier, you know you have to take proper care of your tools and weapons to get the best results when you use them. This is especially true of your knife. Keep it sharpened and ready to use. In a survival situation it is one of your most valuable items.

Can you imagine being in a survival situation without any weapons or tools except your knife? It could happen! You might even be without a knife. So it is important that you know how to improvise.

The main reason why you need a weapon is so you can hunt on the move. A field expedient weapon is not intended to protect you from enemy soldiers, but it can extend your area of defense beyond your fingertips. It can also give you a feeling of security.

Clubs

There are three basic types of clubs.

Simple Club. A simple club is merely a staff or branch short enough for you to swing easily but long enough and strong enough for you to damage whatever you hit.

Weighted Club. A weighted club is any simple club with a weight on one end. The weight may be a natural weight, such as a knot on the wood, or it may be something added, such as a stone. You will also need some type of lashing (page 4-7).

To make a weighted club—

- Find a stone that has a shape that will allow you to lash it securely to the club. A stone with a slight hourglass shape works well.
- Find a piece of wood that is the right length for you. A straight-grained hardwood is best if you can find it.
- Lash the stone to the handle.

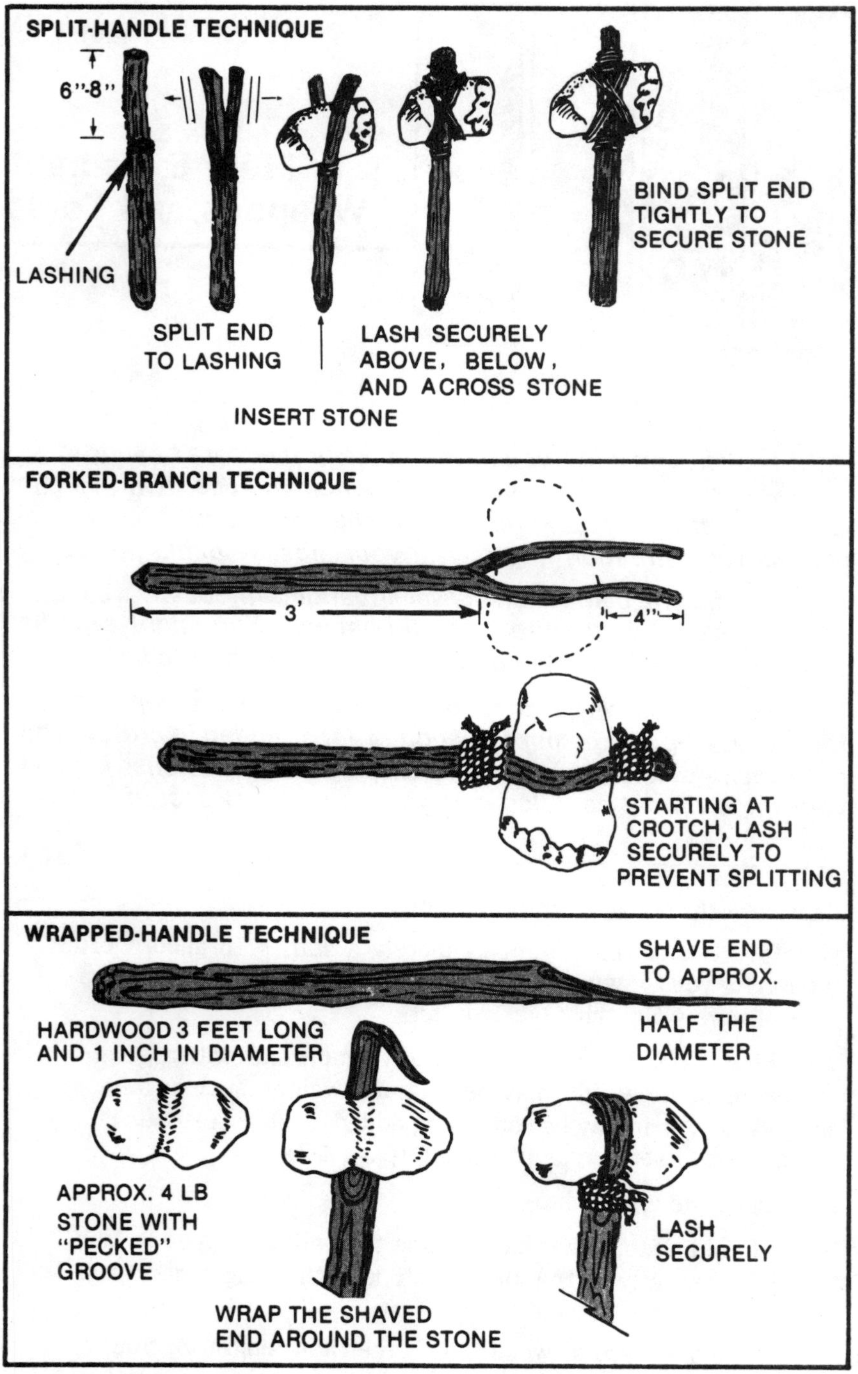

Figure 4-1. Lashing clubs.

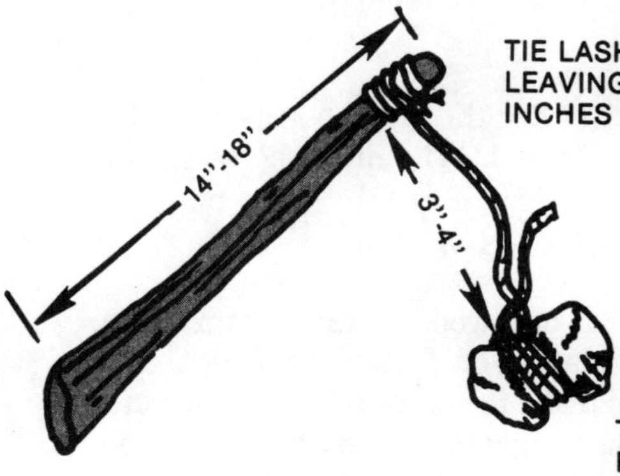

Figure 4-2. Sling club.

There are three techniques for lashing the stone to the handle: split handle, forked branch, and wrapped handle. Which technique you use will depend on the type of handle you use (see figure 4-1).

Sling Club. A sling club is another type of weighted club. However, the weight suspends from the handle (see figure 4-2). A blow from this type of club has more force than a blow from the other types of weighted clubs.

You can make another type of sling club by putting sand or a rock in a sock. This type of weapon, however, is a one-shot deal.

Edged Weapons

In Chapter 2 you learned that you should always have a pocket knife in your survival kit, but maybe you find you need another type knife or a spear. You can use wood, bone, stone, or metal to make a knife or spear blade.

To make a wooden knife—

- Find a straight-grained piece of hardwood about 12 inches long and 1 inch in diameter.
- Shave about 6 inches of this hardwood to form the knife blade, making the point of the blade slightly off- center so that it is not formed by the pith. (The pith makes a weak point.)

- If you have a fire, slowly dry the blade portion over the fire until it is slightly charred. (The drier the wood, the harder the point.)
- Sharpen the blade using a coarse stone.

You can use bamboo instead of hardwood to make your knife. However, after you shape the blade, you should remove only the inside portion of the bamboo to make the blade thinner. The outer layer is harder and should be used to get the hardest possible blade. Char only the inside of the bamboo to harden the blade.

To make a bone knife you will need a suitable bone such as the leg bone of a deer or other medium-size animal, a hard surface and a heavy object with which to shatter the bone, a piece of hardwood for a handle, and lashing material. To do so—

- Find a suitable bone and lay it on a hard surface.
- Hit the bone with a heavy object so that the bone shatters.
- Select a pointed bone splinter that can be further shaped.
- Sharpen the splinter by rubbing it on a rough-textured rock.
- Lash the sharpened bone splinter to a piece of hardwood (handle).

To make a stone knife, you will need a sharp-edged piece of stone, a chipping tool, and a flaking tool. (A chipping tool is a light blunt-edged tool for breaking off small pieces of stone, and a flaking tool is a pointed tool used to break off thin flattened pieces.) You can make a chipping tool from wood, bone, or metal and a flaking tool from bone, antler tine, or soft iron. Lash the blade to the hilt (see figure 4-3).

You can make a metal knife using basic metalworking techniques. To do so—

- Find a piece of soft iron whose shape closely resembles the intended knife (if possible).
- Place the metal on a flat, hard surface and hammer it to get the shape you desire.
- Rub the metal on a rough-textured rock to get a cutting edge and point.
- Lash the knife onto a hardwood handle.

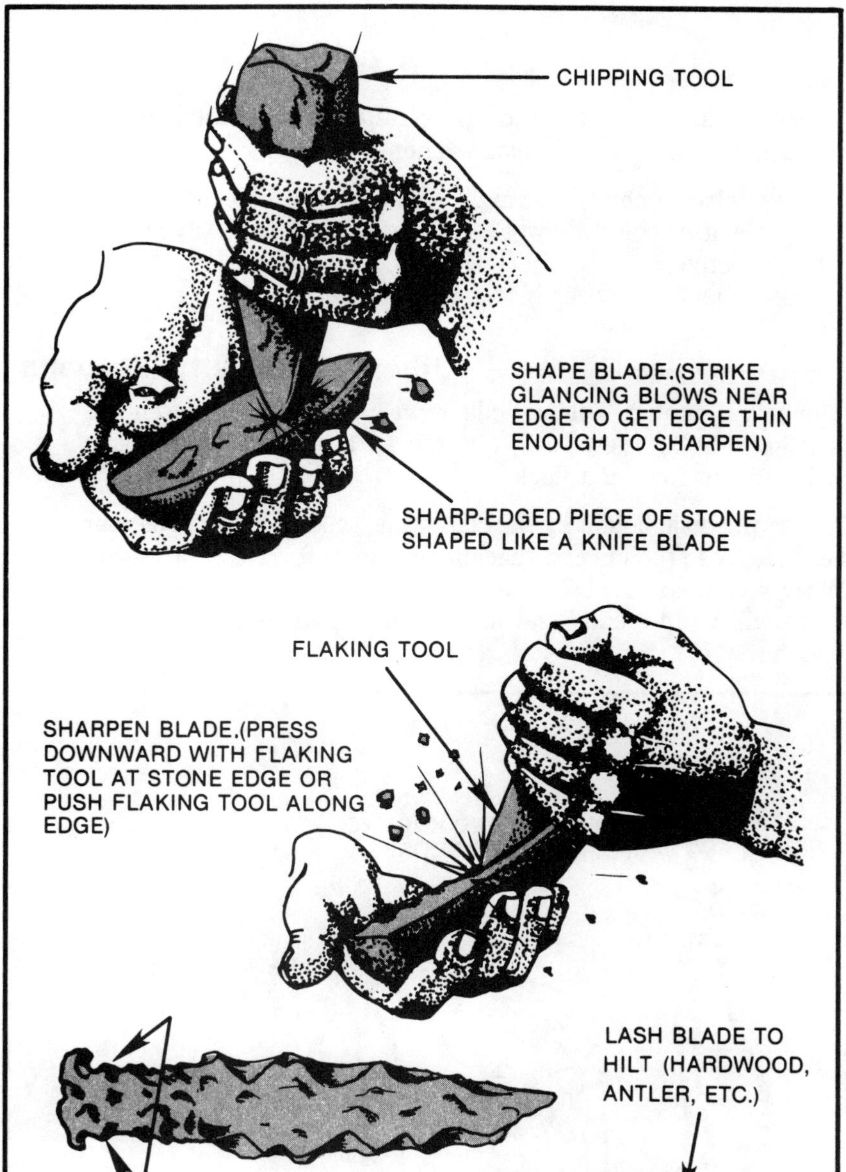

Figure 4-3. Making a stone knife.

To make a spear, use the same procedures to make the blade as you used to make a knife blade. Then—

- Select a shaft (a straight sapling) 4 to 5 feet long. The length should allow you to handle the spear easily and effectively.
- Attach the spear blade to the shaft.

Other Expedient Weapons

The bola is another field expedient weapon that is easy to make (see figure 4-4). It is especially effective for capturing running game or low-flying fowl in a flock.

To use the bola, hold it by the center knot and twirl it above your head. Release the knot so that the bola goes toward your target. When you release the bola, the weighted cords will separate. These cords will wrap around and immobilize the fowl or animal that you hit.

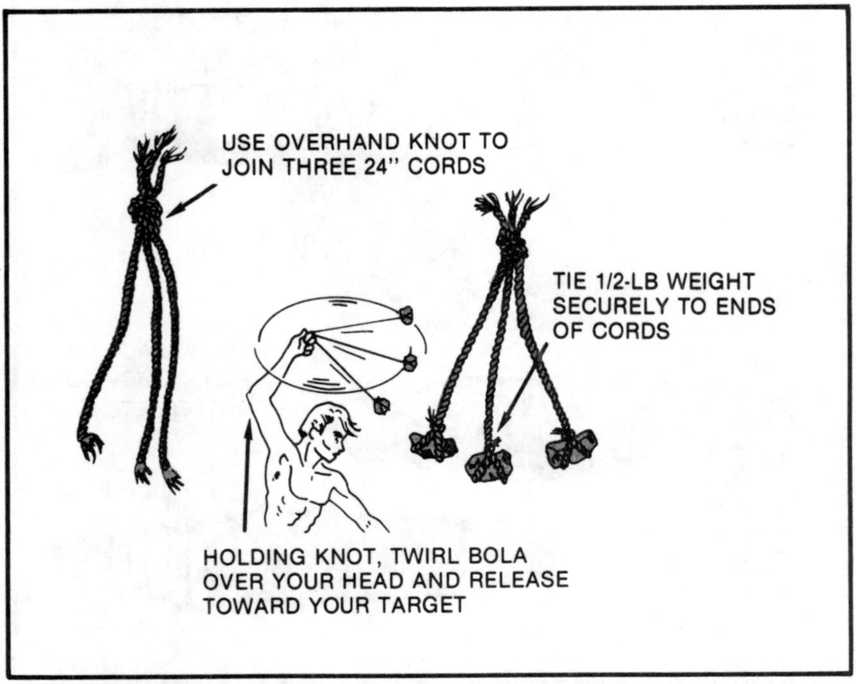

Figure 4-4. Bola.

Figure 4-5. Rabbit stick.

A rabbit stick can be used to kill small game. It is a flat curved club resembling a boomerang. Find a blunt stick that curves naturally at about a 45° angle and shave off two sides so it is flat like a boomerang (see figure 4-5).

Lashing Materials

There are several natural materials you can use to make lashing. However, don't overlook the man-made items you have with you. For instance, you can unravel a cotton web belt and braid enough strands together to give you a cord strong enough to use on a field expedient weapon.

If you have no man-made materials you can use for lashing, look for natural materials. The inner bark of some trees, such as the linden, elm, hickory, white oak, mulberry, chestnut, and red and white cedar, can be shredded and braided together to make cord. After you make the cord, test it to be sure it is strong enough for your purpose.

Sinew (tendon) is the best natural material for lashing small objects. To make the lashing, use tendons from large game, such as deer, as follows:

- Remove the tendons from the game and dry them completely.
- Smash the dried tendons so that they separate into fibers.
- Moisten the fibers and twist them into a continuous strand. If you need stronger lashing material, you can braid the strands.

NOTE: When you use sinew for small lashings, you do not need knots as the moistened sinew is sticky and dries hard.

Rawhide is also good lashing. It is made from the skins of medium or large game. To get the rawhide—

- Skin the animal and remove all fat and meat from the skin.
- Spread out the skin, making sure there are no folds that can trap moisture.
- Cut the dry skin into strips.
- Soak the strips (2 to 4 hours) until they are soft and pliable.

A Drill

On page 9-9 you will find how to make a bow and drill for starting a fire. By adding a drill bit you can use this same drill for boring holes. For the bit you can use a piece of flint or a piece of iron sharpened so it will cut downward and on the side. Drive the bit into the working end of the shaft and lash it securely.

If you need a free hand to hold the material you are drilling, you can fashion a mouthpiece from softwood to use as a cap for the drill. Shape the mouthpiece so you can hold it firmly with your teeth; make the mouthpiece with wing pieces to rest on the outside of the cheeks for comfort. Insert a bone in the mouthpiece at the point where it will connect with the drill and insert the other end of the bone in the drill.

You can also use a hot metal wire or rod to burn a hole in wood.

Chapter 5

Water Procurement

Water is one of your most urgent needs in a survival situation. You can't live long without it, especially in hot areas where you lose so much through sweating. Even in cold areas, you need a minimum of 2 quarts of water a day to maintain efficiency (Chapter 3).

More than three-fourths of your body is composed of fluids. Your body loses fluid as a result of heat, cold, stress, and exertion. The fluid your body loses must be replaced for you to function effectively (Chapter 3). So, one of your first objectives is to obtain an adequate supply of water.

Obtaining Water

Almost any environment has water present to some degree. On pages 5-2 through 5-5 is a figure that lists possible sources of water in various environments. And on page 5-13 is information on how to make the water potable.

NOTE: *If you do not have a canteen, a cup, a can, or other type of water container, you can improvise one from plastic or water-resistant cloth. Shape the plastic or cloth into a bowl by pleating it. Use pins, sharp bones, or other suitable items—even your hands—to hold the pleats.*

If you do not have a reliable source for replenishing your water supply, stay alert for ways in which your environment can help you.

ENVIRONMENT	SOURCE OF WATER	MEANS OF OBTAINING AND/OR MAKING POTABLE	REMARKS
Frigid areas	Snow and ice	Melt and purify	DO NOT EAT without melting! Eating snow and ice can reduce body temperature and will lead to more dehydration. Snow and ice are no purer than the water from which they come. Sea ice that is gray in color or opaque is salty and should not be used without desalting. Sea ice that is crystalline with a bluish cast has little salt in it.
At sea	Sea	Use desalter kit.	DO NOT drink sea water without desalting.
	Rain	Catch rain in tarps or in other water-holding material or containers.	If tarp or water-holding material has become encrusted with salt, wash it in the sea before using (very little salt will remain on it).
	Sea ice		See remarks above for frigid areas.
Beach	Ground	Dig hole deep enough to allow water to seep in; obtain rocks, build fire, and heat rocks; drop hot rocks in water; hold cloth over hole to absorb steam; wring water out of cloth.	Alternate method if helmet or bark pot is available: Fill helmet or pot with sea water; build fire and boil water to produce steam; hold cloth over helmet to absorb steam; wring water out of cloth.

Figure 5-1. Water sources in different environments.

ENVIRONMENT	SOURCE OF WATER	MEANS OF OBTAINING AND/OR MAKING POTABLE	REMARKS
Desert	Ground— • in valleys and low areas. • at foot of concave banks of dry river beds. • at foot of cliffs or rock outcroppings. • at first depression behind first sand dune of dry desert lakes • wherever you find damp surface sand. • wherever you find green vegetation.	Dig holes deep enough to allow water to seep in.	In a sand dune belt, any available water will be found beneath the original valley floor at the edge of the dunes.

Figure 5-1. Water sources in different environments. (Continued)

ENVIRONMENT	SOURCE OF WATER	MEANS OF OBTAINING AND/OR MAKING POTABLE	REMARKS
	Cacti (found only in American deserts except for the barrel cactus, which has been introduced to other countries).	Cut off the top of a barrel cactus; mash or squeeze the pulp. **CAUTION:** *Do not eat pulp. Place pulp in mouth, suck out juice, and discard pulp.*	Without a machete, cutting into a cactus is difficult and takes time since you must get past the long, strong spines and cut through the tough rind.
	Depressions or holes in rock.		Periodic rainfall may accumulate in pools, seep into fissures, or collect in holes in rocks.
	Fissures in rock.	Insert flexible tubing and siphon out water. If fissure is large enough, you can lower a container into it.	
	Porous rock.	Insert flexible tubing and siphon out water.	
	Condensation on metal.	Use cloth to absorb water, then wring water out of cloth.	Extreme temperature variations between night and day may cause condensation on metal surfaces.

Figure 5-1. Water sources in different environments. (Continued)

ENVIRONMENT	SOURCE OF WATER	MEANS OF OBTAINING AND/OR MAKING POTABLE	REMARKS
			NOTE: Following are "signs" to watch for in the desert to help you find water: (1) All trails lead to water. You should follow in the direction in which the trails converge. Trails may be marked by signs of camps—campfire ashes, animal droppings, and trampled terrain.  (2) Flocks of birds will circle over waterholes. Some birds fly to waterholes at dawn and sunset. Their flight at these times is generally fast and close to the ground. Bird tracks or chirping sounds in the evening or early morning sometimes indicate that water is nearby.

Figure 5-1. Water sources in different environments. (Continued)

FLUID	REMARKS
Alcoholic beverages	Dehydrates the body and clouds judgment.
Urine	Contains harmful body wastes. Is about 2 percent salt.
Blood	Is salty and is considered a food; therefore, requires additional body fluids to digest.
Sea water	Is about 4 percent salt. It takes about 2 quarts of body fluids to rid the body of waste from 1 quart of sea water. Therefore, by drinking sea water you are depleting your body's water supply, which can cause death.

Figure 5-2. The effects of substitute fluids.

CAUTION: *Do not substitue the fluids listed in figure 5-2 for water.*

Heavy dew can provide water. Tie rags or tufts of fine grass around your ankles and walk through dew-covered grass before sunrise. As the rags or grass tufts absorb the dew, wring the water out into a container. Repeat the process until you have a supply of water or until the dew is gone. Australian natives sometimes mop up as much as a quart an hour in this way.

Bees or ants going into a hole in a tree may indicate a water-filled hole. Siphon the water out with plastic tubing, scoop it out with an improvised dipper, or stuff cloth in the hole to absorb the water, then wring it out.

Water sometimes gathers in tree crotches or rock crevices. Use the same procedure as above to get the water.

In arid areas, bird droppings around a crack in the rocks may indicate water in or near the crack.

Green bamboo thickets are an excellent source of fresh water. Water from green bamboo is clear and without odor. To get the water, bend over a stalk of green bamboo, tie it down, and cut off the top (figure 5-3). The water will drip freely during the night.

Old, cracked bamboo segments may contain water.

CAUTION: *You must purify the water before drinking it.*

Figure 5-3. Getting water from green bamboo.

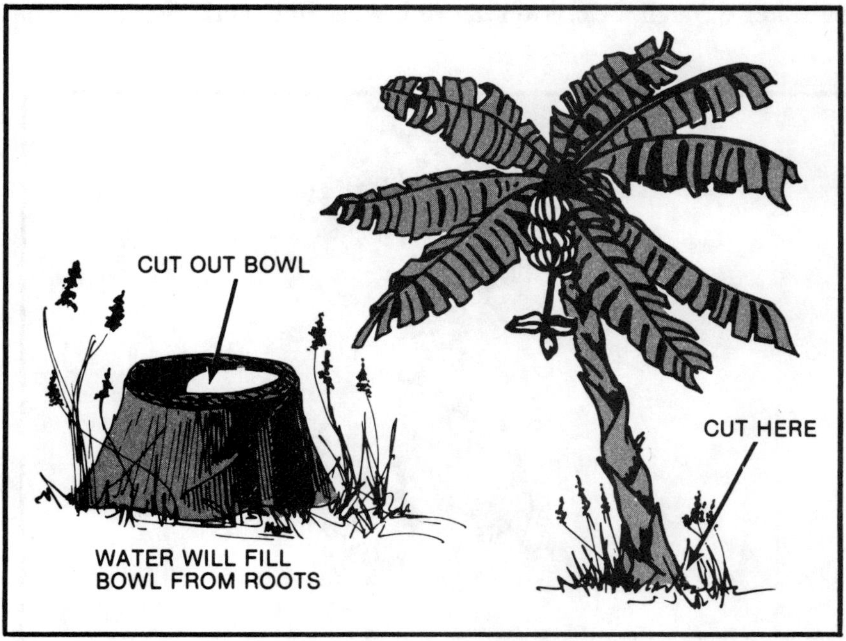

Figure 5-4. Water from plantain or banana tree stump.

Wherever you find banana or plantain trees, you can get water. Cut off the tree, leaving about a 1-foot stump, and scoop out the center of the stump so that the hollow is bowl-shaped. Water from the roots will immediately start to fill the hollow. The first three fillings of water will be bitter, but succeeding fillings will be palatable. The stump (figure 5-4) will supply water for up to 4 days. Be sure to cover it to keep out insects.

Some tropical vines can give you water. Cut a notch in the vine as high as you can reach, then cut the vine off close to the ground. Catch the dropping liquid in a container or in your mouth (figure 5-5). *Do not* drink the liquid if it is sticky, milky, or bitter tasting.

The milk from green (unripe) coconuts is a good thirst quencher. However, the milk from mature coconuts contains an oil that acts as a laxative. Drink in moderation only.

In the American tropics you may find large trees whose branches support air plants. These air plants may hold a considerable amount of rainwater in their overlapping, thickly growing leaves. Strain the water through a cloth to remove insects and debris.

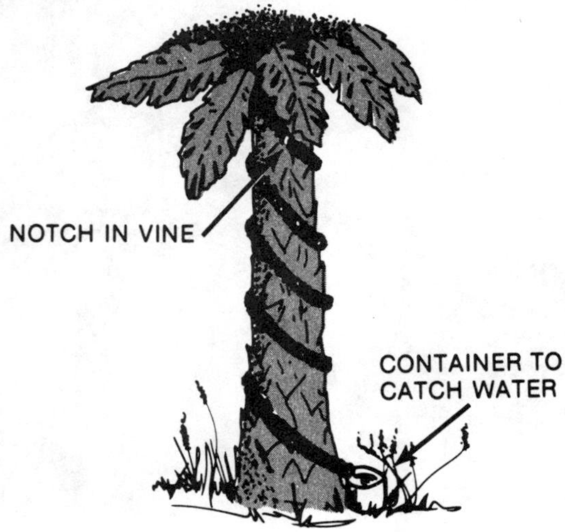

Figure 5-5. Getting water from a vine.

You can get water from plants with moist pulpy centers. Cut off a section of plant and squeeze or smash the pulp so that the moisture runs out. Catch the liquid in a container.

Plant roots may provide water. Dig or pry the roots out of the ground, cut them into short pieces, and remove the bark. Usually you can suck water from the roots.

Fleshy leaves, stems, or stalks, such as bamboo contain water. Cut or notch stalks at the base of a joint to drain out the liquid.

The following trees can also provide water:

- Palms, such as the buri, coconut, sugar, and nipa, contain liquid. Bruise a lower frond and pull it down so the tree will "bleed" at the injury.
- The traveler's tree of Madagascar has a cup-like sheath at the base of its leaves in which water collects.
- The leaf bases and roots of the umbrella tree of western tropical Africa can provide water.
- The baobab tree of the sandy plains of northern Australia and Africa collects water in its bottle-like trunk during the wet season. Frequently, clear fresh water can be found in these trees after weeks of dry weather.

CAUTION: Do not keep sap from plants longer than 24 hours. It begins fermenting, becoming dangerous as a water source.

Stills can be used in various areas of the world. They draw moisture from the ground and from plant material. But you need certain materials to build a still, and you need time to let it collect the water. It takes about 24 hours to obtain from 1 pint to 1 quart of water.

Constructing a Still

There are two types of water stills: the aboveground and the below-ground.

For the aboveground still, see figure 5-6 .

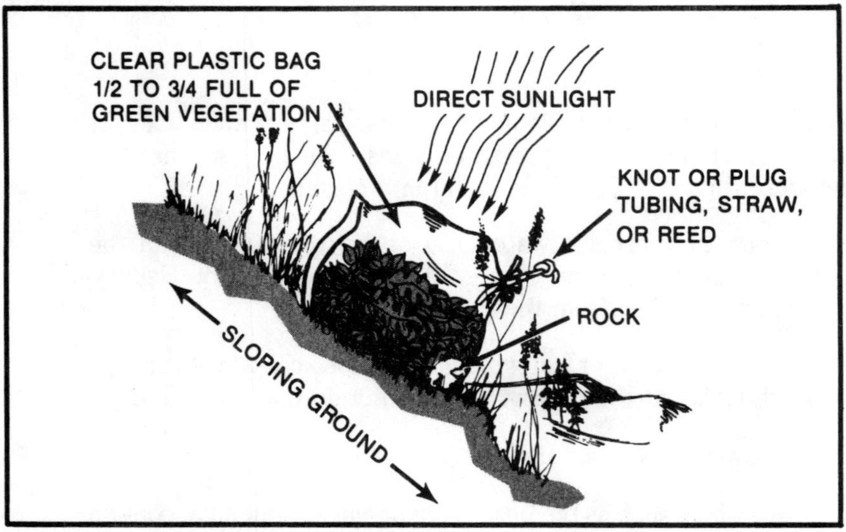

Figure 5-6. Aboveground solar water still.

You must have a sunny slope on which to place the still. Proceed as follows:

- Fill the bag with air by turning the opening into the breeze or by "scooping" air into the bag.
- Fill the plastic bag half to three-fourths full of green leafy vegetation. Be sure to remove all hard sticks or sharp spines that might puncture the bag. *CAUTION: Do not use poisonous vegetation. It will provide poisonous liquid.*
- Place a small rock or similar item into the bag.
- Close the bag and tie the mouth securely as close to the end of the bag as possible to retain the maximum amount of air space. (If you have a piece of tubing, small straw, or hollow reed, insert one end in the mouth of the bag before you tie it securely. Then tie off or plug the tubing so that air will not escape. This will allow you to drain out condensed water without untying the bag).
- Place the bag, mouth downhill, on a slope in full sunlight. The mouth of the bag should be slightly higher than the low point in the bag.
- Settle the bag in place so that the rock works itself into the low point in the bag.

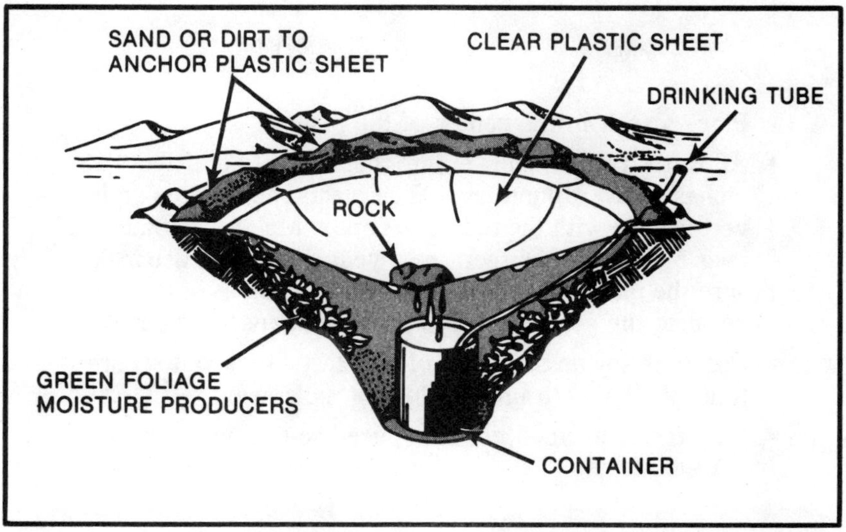

Figure 5-7. Belowground still.

To get the condensed water from the still, loosen the tie around the bag's mouth and tip the bag so that the water collected around the rock will drain out. Then retie the mouth securely and reposition the still to allow further condensation.

For the belowground still (figure 5-7), you will need a digging tool.

You should select a site where you believe the soil will contain moisture (such as a dry stream bed or a low spot where rainwater has collected), where the soil will be easy to dig, and where sunlight hits most of the day. Proceed as follows:

- Dig a bowl-shaped hole approximately 3 feet across and 2 feet deep.
- Dig a sump in the center of the hole. The depth and the perimeter of the sump will depend on the size of the container that you have to set in it. The bottom of the sump should allow the container to stand upright.
- Anchor the tubing to the bottom of the container by forming a loose overhand knot in the tubing.
- Place the container upright in the sump.
- Extend the unanchored end of the tubing up, over, and beyond the lip of the hole.
- Place the plastic sheeting over the hole, covering the edges with soil to hold it in place.

- Place a rock in the center of the plastic.
- Allow the plastic to lower into the hole until it is about 15 inches below ground level. The plastic now forms an inverted cone with the rock at its apex. Make sure that the apex of the cone is directly over your container. Also make sure the plastic cone does not touch the sides of the hole because the earth will absorb the condensed water.
- Put more soil on the edges of the plastic to hold it securely in place and to prevent loss of moisture.
- Plug the tube when not being used so that moisture will not evaporate.

You can drink water without disturbing the still by using the tube as a straw.

You may want to use plants in the hole as a moisture source. If so, when you dig the hole you should dig out additional soil from the sides of the hole to form a slope on which to place the plants. Then proceed as above.

If polluted water is your only moisture source, dig a small trough outside the hole about 10 inches away from the lip of your still (figure 5-8). Dig the trough about 10 inches deep and 3 inches wide. Pour the polluted water in the trough. Be sure you do not spill any polluted water around the rim of the hole where the plastic touches the soil. The purpose of the trough is to hold the polluted water so that the soil will filter it as it is drawn into the still. The water then condenses on the plastic and drains into the container. This process works extremely well when your only water source is saltwater.

Making Water Potable

Rainwater collected in clean containers or in plants is generally safe for drinking. However, you *must* purify water from lakes, ponds, swamps, springs, or streams, especially those near human habitation or in the tropics. When at all possible, you must disinfect all water obtained from vegetation or from the ground by using iodine or chlorine or by boiling.

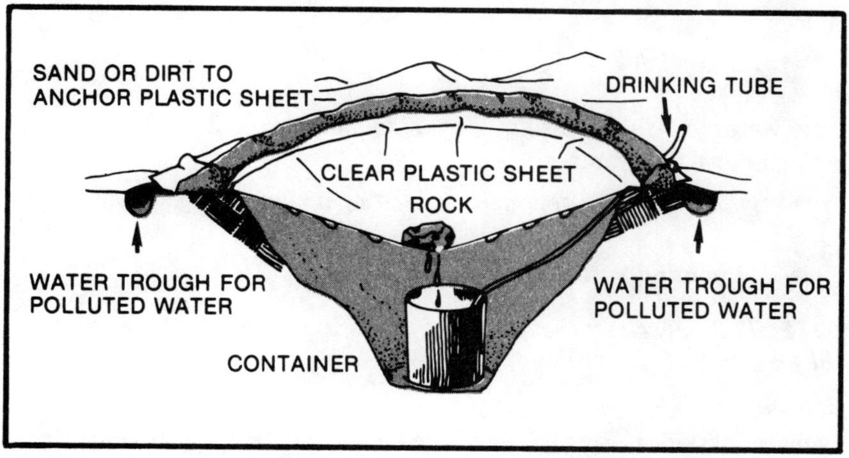

Figure 5-8. Belowground still for obtaining potable water from polluted water.

You can purify water by—

- Using water purification tablets. (Follow the directions provided.)
- Pouring 5 drops of 2 percent tincture of iodine in a canteen full of clear water, and 10 drops in a canteen full of cloudy or cold water. (Let the canteen of water stand for 30 minutes before drinking.)
- Boiling water for 1 minute at sea level, adding 1 minute for each additional 1,000 feet above sea level, or boil *for 10 minutes* no matter where you are.

By drinking nonpotable water you may contract diseases or swallow organisms that can harm you. For example—

- Dysentery (severe, prolonged diarrhea with bloody stools, fever, and weakness).
- Cholera and typhoid. (You may be susceptible even though you have been innoculated.)
- Flukes. (Stagnant, polluted water—especially in tropical areas— often contains blood flukes. If you swallow flukes, they will bore into the bloodstream, live as parasites, and cause disease).
- Leeches. (If you swallow a leech, it can hook onto the throat passage or inside the nose. It will suck blood, create a wound, and move to another area. Each bleeding wound may become infected).

If the water you find is also muddy, stagnant, and foul smelling, you can clear the water—

- By placing it in a container and letting it stand for 12 hours or
- By pouring it through a filtering system.

NOTE: *This only clears the water and makes it more palatable. You will have to purify it.*

To make a filtering system, place several inches or layers of filtering material such as sand, crushed rock, charcoal, or cloth in bamboo, a hollow log, or an article of clothing (figure 5-9).

Remove the odor from water by adding charcoal from your fire to it. Let the water stand for 45 minutes before drinking it.

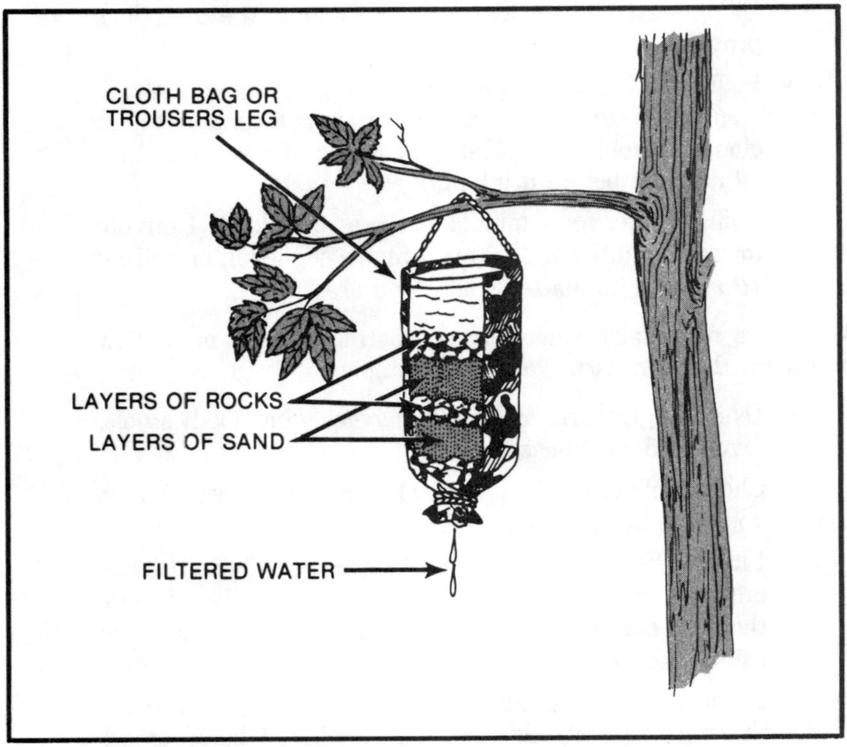

Figure 5-9. Water filtering system.

Chapter 6

Wild Plants For Food

After water, food is your most urgent need. So in a survival situation you should always be on the lookout for wild foods and live off the land whenever possible. Save your rations, if any, for emergencies.

You should keep in mind, however, that—

- *You can live many days without food if you have water. So when water is no problem, drink more than your normal amount to keep fit.*
- *Eating increases thirst. If you have less than a quart of water daily, avoid dry, starchy, and highly flavored foods and meat. Eat foods that have a high carbohydrate content such as hard candy and fruit bars if available.*
- *Every bit of work requires additional food and water, so keep work to a minimum when you have limited food and water.*
- *With few exceptions, everything that grows from the soil or that walks, crawls, or swims is a potential food source.*

Nature can provide food that will enable you to survive. You must know, however, how to find and obtain the food. In this chapter you will learn about obtaining and preparing plant food. In Chapter 7 you will learn how to obtain and prepare wildlife for food.

Nourishment From Plants

Plants are a valuable food source. Although they may not provide a balanced diet, they will sustain you even in the arctic where the heat-producing qualities of meat are normally essential. Many plant foods such as nuts and seeds will give you enough protein for normal efficiency. Roots, green vegetables, and plant food containing natural sugar will provide calories and carbohydrates that give the body energy.

The food value of plants becomes more and more important if you are eluding the enemy or if you are in an area where wildlife is scarce. For instance—

- You can dry plants by wind, air, sun, or fire. This retards spoilage so that you can store or carry the plant food with you to use when needed.
- You can obtain plants more easily and more quietly than meat. This is extremely important when the enemy is near.

Edibility of Plants

Being able to recognize both cultivated and wild edible plants is important in a survival situation. However, since information on cultivated plants is readily available, most of the information in this manual is about wild plants.

There are certain things you should keep in mind when collecting edible plants:

- Cultivated plants and wild plants growing in or near cultivated plants may have been sprayed with pesticides. So thoroughly wash whatever plants you collect.
- The surface of any plant food that grows or is washed in contaminated water is also contaminated. If you are going to eat the plant raw, wash it in water suitable for drinking.
- Some plants may have fungal toxins that are extremely poisonous. To lessen the chances that these toxins are present, collect fresh seeds, fruit, or leaves—not those that have fallen to the ground.
- Plants of the same species may differ in the amount of toxic or subtoxic compounds they contain because of different environmental and genetic factors. One example of this is the foliage of the common chokeberry. Some chokeberry

plants have high concentrations of cyanide compounds, other plants have low concentrations.

- Some people are more susceptible than others to gastric upset from plants. If you are sensitive this way, avoid unknown wild plants. If you are extremely sensitive to poison ivy, avoid products from this family, including drinks made from sumacs, mangos, and cashews.
- There are some edible wild plants, such as acorns and water lily rhizomes, that are bitter. These bitter substances (usually tannin compounds) make them unpalatable. Boiling in several changes of water will help remove these substances.
- There are many valuable wild plants that have high concentrations of oxalate compounds. Oxalates usually produce a sharp burning sensation in your mouth. And they are bad for the kidneys. Boiling usually destroys these oxalates.
- The only way to tell if a mushroom is edible is by proper determination. Even then, some species are questionable. So do not eat mushrooms.

There are many, many plants throughout the world. Tasting or swallowing even a small portion of some can cause severe discomfort, extreme internal disorders, or death. Therefore, if you have the slightest doubt as to the edibility of a plant, apply the Universal Edibility Test (figure 6-1) before eating any part of it.

Before testing a plant for edibility, make sure there are a sufficient number of the plants to make testing worth your time and effort. You need more than 24 hours to apply the edibility test.

Keep in mind that eating large amounts of plant food on an empty stomach may cause diarrhea or cramps. Two good examples of familiar foods that can cause this problem are green apples and too many fresh berries. Even though you have tested plant food and found it safe, eat it in moderation with other foods.

You can see from the steps and time involved in testing edibility just how important it is to be able to identify edible plants.

1. Test only one part of a potential food plant at a time.
2. Break the plant into its basic components—leaves, stems, roots, buds, and flowers.
3. Smell the food for strong or acid odors. Keep in mind that smell alone does not indicate a plant is inedible.
4. Do not eat for 8 hours before starting the test.
5. During the 8 hours you are abstaining from eating, test for contact poisoning by placing a piece of the plant part you are testing on the inside of your elbow or wrist. Usually 15 minutes is enough time to allow for a reaction.
6. During the test period, take nothing by mouth except purified water and the plant part being tested.
7. Select a small portion of a single component and prepare it the way you plan to eat it.
8. Before putting the prepared plant part in your mouth, touch a small portion (a pinch) to the outer surface of the lip to test for burning or itching.
9. If after 3 minutes there is no reaction on your lip, place the plant part on your tongue, holding it there for 15 minutes.
10. If there is no reaction, thoroughly chew a pinch and hold it in your mouth for 15 minutes. DO NOT SWALLOW.
11. If no burning, itching, numbing, stinging, or other irritation occurs during the 15 minutes, swallow the food.
12. Wait 8 hours. If any ill effects occur during this period, induce vomiting and drink a lot of water.
13. If no ill effects occur, eat ½ cup of the same plant part prepared the same way. Wait another 8 hours. If no ill effects occur, the plant part as prepared is safe for eating.

CAUTION: Test all parts of the plant for edibility, as some plants have both edible and inedible parts. Do not assume that a part that proved edible when cooked is also edible when raw. Test the part raw to ensure edibility before eating raw.

Figure 6-1. Universal Edibility Test.

Do not eat unknown plants that—

- Have a milky sap or a sap that turns black when exposed to air.
- Are mushroomlike.
- Resemble onion or garlic.
- Resemble parsley, parsnip, or dill.
- Have carrotlike leaves, roots, or tubers.

Preparation of Plant Food

Although some plants or plant parts are edible raw, others must be cooked to be edible or palatable.

Some methods of improving the taste of plant food are soaking, parboiling, cooking, or leaching. (Leaching is done by crushing food, placing it in some sort of strainer, and pouring boiling water through it.)

Leaves, Stems, and Buds. Boil until tender. Several changes of water help to eliminate bitterness.

Roots and Tubers. Boil, bake, or roast. Boiling removes harmful substances such as oxalic acid crystals.

Nuts. Leach or soak acorns in water to remove the bitterness. Although chestnuts are edible raw, they are tastier roasted or steamed.

Grains and Seeds. Parch to improve the taste, or grind into meal to use as a thickener with soups or stews or to use as flour to make bread.

Sap. If the sap contains sugar, dehydrate it by boiling until the water in it is gone.

Fruit. Bake or roast tough, heavy-skinned fruit. Boil juicy fruit.

Cooking Methods

In a survival situation, you may have to improvise containers for cooking your food. The cooking processes, however, are the same as you would use in a kitchen. They are boiling, frying, parching, baking, steaming, and roasting or broiling.

Boiling is one of the best methods to use in that you can retain the juices that contain salts and nutrients.

If you do not have a metal container in which to boil your food, use a rock that has a hole in it. Or use a hollowed out piece of wood that will hold your food and enough water to cook it in. Hang the wooden container over the fire and add hot rocks to the water and food; remove the rocks as they cool and add more hot rocks until your food is cooked.

CAUTION: *Do not use rocks with a high moisture content, such as those from streams or wet areas, as they may explode.*

You can also use this method with containers made of bark or leaves, but these containers will burn above the water line unless you keep them moist or keep the fire low.

Other items you can use as containers for boiling food are coconut shells, sea shells, turtle shells, half sections of bamboo, complete sections of bamboo (figure 6-2), even a bag made from the stomach or skin of an animal.

CAUTION: *A closed section of bamboo may explode when heated.*

Rock frying is a method you can use when you do not have a frying pan. Place a flat or slightly concave rock on the fire. When the rock is hot, place grease and your food on it and heat until cooked.

Parching works especially well with nuts and grains. Place the nuts or grain in a container or on a rock and heat slowly until the nuts or grain is scorched.

Baking is cooking in an oven over slow, steady, moderate heat. An oven may be improvised using a pit under a fire, a closed container, or a wrapping of leaves or clay.

One method is to dig a pit and partly fill it with hot coals. Put your food with some water in a covered container. Place the covered container in the pit. Cover the container with a layer of coals and thin layer of dirt.

Another pit method is to line the pit with moisture-free stones. Build a fire in the pit, and as the coals burn down, scrape them back, put the covered container in, and continue as above.

Figure 6-2. Containers for boiling food.

Figure 6-3. Steaming food.

A rock-type oven is also suitable for baking. Use a rock for the bottom. Use four more rocks as the walls of the oven and a rock large enough to cover your rock "box" as the top. Place coals inside around the sides of the bottom rock and then place the food on the bottom rock. By adding or taking away coals, you can keep the heat relatively uniform.

Steaming is also done with a pit. It works best with shellfish or other foods that require little cooking such as plantains or green bananas. Wrap the food in large leaves or moss. Place one piece or one layer of wrapped food in the coals-lined pit. Add another layer of leaves or moss. Continue alternating layers of wrapped food and layers of leaves or moss until the pit is almost full. Push a stick down through the layers of food and leaves or moss. Try to seal the pit with more leaves or dirt, and then remove the stick. This is a slow but effective way to cook.

Roasting or broiling is done with a skewer or spit. You can cook all meats over an open fire using this method. Simply stick the meat on a nontoxic spit and place it over the fire until done. This method sears the flesh, which holds the juices in and allows the meat to cook in its own juice. This is a good method for cooking whole fowls or small animals.

Chapter 7

Wildlife For Food

Meat is more nourishing than plant food. In fact, it may even be more readily available in some places. But to get meat, you need to know the habits of, and how to capture,. the various wildlife.

To satisfy your immediate food needs, you should seek first the more abundant and easier to obtain wildlife, such as insects, crustaceans, mollusks, fish, and reptiles.

Insects

Many insects that are abundant in various parts of the world are a valuable source of food, and most are easily caught by hand. These include large grubs (the larvae of insects), locusts, grasshoppers, ants, and termites. You can fry, boil, or roast them, but you may prefer to add them to a stew containing other foods to make them more palatable. You can even eat them raw except for grasshoppers, which may contain harmful parasites. (Do not try to eat the large grasshopper legs as they have barbs that may stick in your throat.) Insects are high in fat content; many, ounce for ounce, have higher protein value than beef.

Woodgrubs are found in rotten logs, in the ground, and under the bark of dead trees. Locusts, grasshoppers, and ants are found throughout temperate and tropical zones. Termites are generally available in jungles.

Water Wildlife

You should never go hungry when you're near a body of water—a lake, a stream, a river, an ocean. Most bodies of freshwater and saltwater contain crustaceans, mollusks, and fish as well as other forms of edible water creatures. And you can usually catch water wildlife faster, more easily, more quietly, and in greater quantity than you can land wildlife. You need only know their habits, the best time of day to catch them, and how to catch them to have a plentiful supply of food.

Crustaceans. This class includes freshwater and saltwater crabs, crayfish, lobsters, shrimp, and prawns. All are edible, but you should always cook the freshwater crustaceans as they may harbor harmful parasites.

Freshwater shrimp are abundant in tropical streams, especially where the water is sluggish. Look for them swimming or clinging to branches or vegetation in the water.

Saltwater shrimp live on or near the sea bottom. You can scrap them up, or at night you can lure them to the surface with a light and catch them with a hand net.

Freshwater crabs and crayfish are found on moss beds under rocks and brush in streams or swimming in shallow water. You can pick them up by hand or scoop them up with a dip net.

Saltwater crayfish and lobsters are found on the ocean bottom in water 10 to 30 feet deep. You can use lobster traps (page 7-11), a jug, or a baited hook to catch them. Lift your catch out of the water with a dip net.

Many species of crabs and lobsters are nocturnal and are most easily caught at night.

Crabs creep, climb, and burrow. You can easily catch them in shallow water using a dip net. You can also catch them in traps baited with fish heads or animal guts.

Mollusks. This class includes freshwater and saltwater shellfish such as snails, clams, mussels, bivalves, periwinkles, chitons, and sea urchins. Bivalves similar to our freshwater mussel and terrestrial and aquatic snails are found worldwide under all water conditions.

River snails or freshwater periwinkles are plentiful in the rivers, streams, and lakes of northern coniferous forests. These snails may be pencilpoint or globular in shape.

In freshwater, look for mollusks in the shallows, especially in water with a sandy or muddy bottom. Look for the narrow trails they leave in the mud or for the dark elliptical slit of their open valves.

Near the sea, wait for low tide and then check in the tidal pools and the wet sand. Rocks along beaches or extending as reefs into deeper water often bear clinging shellfish. Snails and limpets cling to rocks and seaweed from the low-water mark up. Large snails called chitons adhere tightly to rocks above the surf line.

Mussels usually form dense colonies in rock pools, on logs, or at the base of boulders.

CAUTION: Mussels are poisonous in tropical zones during the summer.

Mollusks should be steamed, boiled, or baked in the shell. They make excellent stews in combination with greens and tubers.

CAUTION: Do not eat shellfish that are not covered by water at high tide.

Fish. Of the wildlife around or in freshwater, fish are probably the most difficult to catch. But you can catch fish, even without modern fishing equipment, if you know when, where, and how to fish. And you can easily make hooks and line and find bait in most areas near water.

How to Make Fishhooks. You can make hooks from pins, needles, wire, small nails, or any piece of metal; out of wood, coconut shell, bone, thorns, flint, seashell, tortoise shell; or out of a combination of these items. One way to make a wooden fishhook (figure 7-1) is as follows:

- Cut a piece of hardwood about 2 inches long and ½ inch in diameter to form the shank.
- Cut a notch near one end of the wood in which to place the point.
- Place the point (a small, pointed piece of wood, a pin, nail, piece of bone, or similar item) in the notch.
- Hold the point in the notch and tie it securely.

To tie securely—

- Make a clove hitch in the cord and take about six turns around the end of the point and the hardwood shank,
- Take two or three frapping turns between the point and the shank to spread and hold the point firmly angled to the shank.

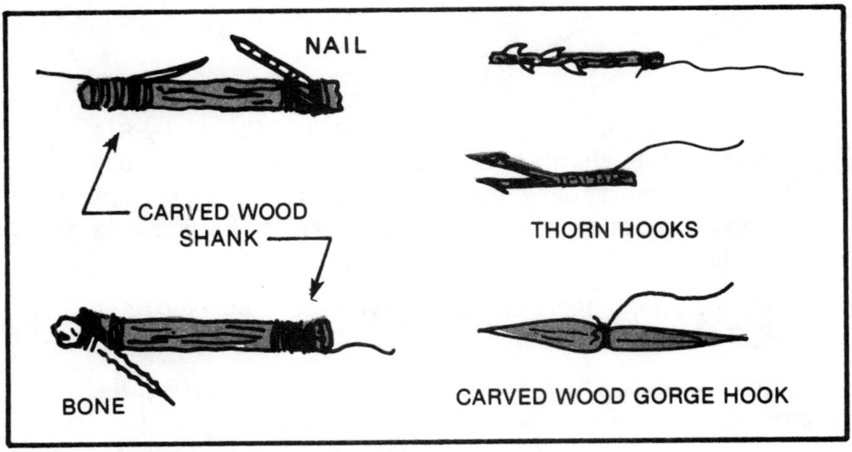

Figure 7-1. Improvised fishhooks.

- Take two more turns around the shank and tie with a clove hitch.

Make sure that the fishhook is the right size for the size of fish found in the area.

How to Make Fishlines. You can make a fishline from suspension lines or from plant or cloth fibers. The inner bark of trees make the best fibers to use. To make a line from the fibers (figure 7-2)—

- Knot the ends of two strands and secure them to a solid base.
- Hold a strand in each hand and twist clockwise.
- Cross one twisted strand counterclockwise over the other.
- Add fiber as necessary to increase the length of the line.

NOTE: *Fibers from hemp, nettle, common and swamp milkweed, yucca, and reeds also make good lines.*

How to Find Bait. Generally, fish bite bait that is native to their area. So look in the water near the shores for crabs, fish eggs, and minnows and on the bank for worms and insects. After you catch a fish, open it and examine its stomach and intestines to see what it was eating. Try to duplicate its food. You can use feathers, pieces of brightly colored cloth, or bits of bright metal or shell to make artificial lures.

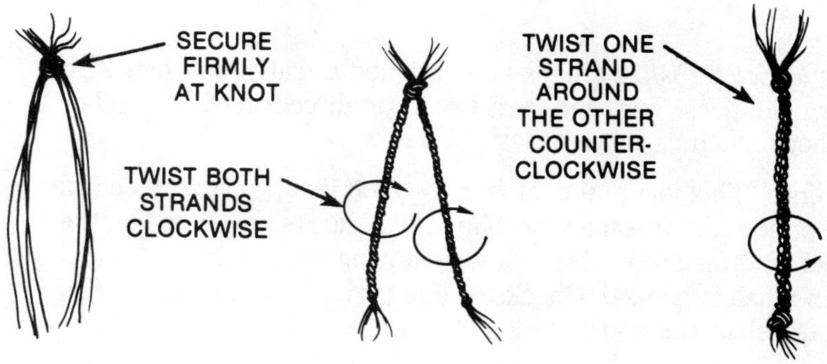

Figure 7-2. Making lines from plant fibers.

When to Fish. As a general rule, look for fish to feed just before dawn and just after dusk, just before a storm as the front is moving in, and at night when the moon is full or waning. Rising fish and jumping minnows are often signs of feeding fish.

Where to Fish. The body of water, the zone in which it is located, the time of year, and the time of day all have a bearing on where to fish.

In lakes or large streams, fish tend to approach the banks and shallows in the morning and evening.

In streams, fish often gather in pools and deep calm water, at the bottom of riffles and small rapids, at the tail of a pool, in eddies below rocks or logs, under deep undercut banks, in the shade of overhanging bushes, and around submerged logs and rocks.

Fish seek shelter at the mouths of small tributary streams when the main rivers or streams are high or muddy.

In shallow streams during hot weather, fish gather in the deepest pools, in places where cool underground water enters the main stream, and under rocks.

In temperate zones in cool spring weather, fish tend to move to shallow water that is warmed by the sun.

How to Catch Fish. Perhaps you've failed to catch any fish with your pole, hook, line, and bait. Don't get discouraged. Try other methods, such as those described below.

Set lines. This is a practical way to catch fish if you stay near a lake or stream for some time. Tie several hooks on a line, bait the hooks, and fasten the line to a low-hanging branch that will bend when a fish is hooked. Check the line periodically to remove fish and to rebait the hooks.

A gorge or skewer hook is an excellent hook for a set line. To make a gorge hook (figure 7-1), sharpen both ends of a short piece of bone or wood and cut a notch circling the middle of it. Tie a line to it. Bait the hook so that when the skewer is in the water, it will lie back along the line. Fix the line in the water. When a fish swallows the bait, the line sets and the gorge swings crosswise and lodges in the fish's gullet or stomach.

Set a stakeout. This is a fishing device you can use secretly. You can set many stakeouts with little danger that other persons will detect them. To make a stakeout (figure 7-3), stick two reeds on the bottom of the stream or lake. Run a line between the two reeds (the line can slide up and down the reeds). To this line, tie two other lines with fishhooks attached. Make sure the lines will not become entangled with each other or with the upright reeds.

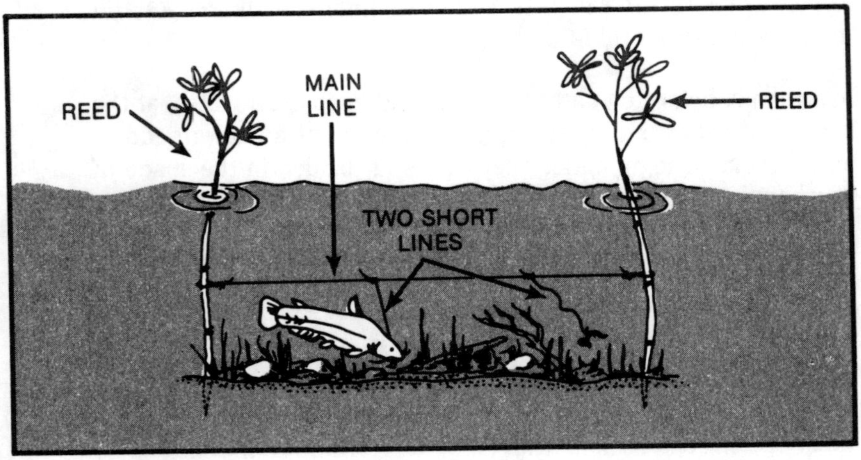

Figure 7-3. Stakeout.

Just before dark, bait hooks with worms, bee larva, or other suitable bait. At dark, lower lines in the water. Wait 1 to 1-1/2 hours before checking the lines for fish; rebait the hooks if needed and then wait another hour before checking for fish. Check the lines again at first light.

Try jigging. This method is especially effective at night. You will need a limber cane or pole 8 to 10 feet long; a piece of line about 10 inches long; a hook; a piece of bright metal shaped like a commercial fishing spoon; and a small strip of white meat or fish intestine. Tie the line to the end of the pole. Attach the spoon and the hook on the line so that the hook is just below the spoon. Bait the hook and dabble it and the spoon just below the surface of the water near lily pads or weed beds. Occasionally slap the water with the tip of the pole to attract large fish to the bait.

Use your hands. This method is effective in small streams with undercut banks or in shallow ponds left by receding flood waters. Place your hands in the water and slowly reach under the bank, keeping your hands close to the bottom if possible. Move your fingers slightly until you contact a fish. Then work your hand gently along its belly until you reach its gills. Grasp the fish firmly just behind the gills. If you are handling catfish or fish with spiney dorsal fins, take care to avoid getting stuck.

Try muddying. Small pools caused by receding waters of flooded streams often contain many fish. By stamping the bottoms of these pools or stirring the mud with a stick, you make the water muddy so that the fish rise to the surface to seek clearer water. You can then club the fish or toss them out of the water with your hands.

Use a net. A net is more efficient than the other means discussed above for catching fish. But it takes time to make a large net. You can, however, quickly make a dip net to catch small fish to use for bait or to eat. Fish too small to hook or spear are usually abundant at the edges of lakes and streams or in their tributaries. A dip net to catch these fish can be made using a forked sapling and a piece of cloth, such as your undershirt. Bend the two limbs of the sapling fork, securing the ends firmly together, so that they form a circular frame. Close the neck and armholes of the undershirt by making a knot in the top. Attach the bottom of the undershirt to the circular frame, using pins, wire, or any other available item that will secure it.

An even better way to catch fish is with a gill net. But you need time to make it. You also need a suspension line or similar line and two uprights or trees standing apart about the same distance as the planned length of your net. To make it (figure 7-4)—

- Remove the core lines from your suspension lines.
- Suspend the suspension line casing at about eye level between two trees. (We'll call this the main suspension line.) The length of this line depends on how long you wish to make the net and how much line you have.
- Attach an even number of core lines to the main suspension line using a Prusik knot or girth hitch. (The double lines from these knots should be about the same length.) The number of core lines used and their distance apart are determined by the length of the net and the mesh size desired. Use 1-inch space to make 1-inch mesh, 2-inch space to make 2-inch mesh, and so forth.
- Starting at one end of the main suspension line, tie the second and third core lines together with an overhand knot, fourth and fifth lines together, and so forth. You will have one line remaining at the end.
- For the second row of knots, tie the first and second core lines together, the third and fourth together, and so forth.

NOTE: *You can use a guide line, moving it down for each row of knots, to make the mesh the same size. Be sure to place the guide line on the side of the net opposite to you so it won't interfere with your work.*

Follow the same procedure for the third row as you did for the first row and the same procedure for the fourth as for the second, and so forth.

Continue making the mesh until you reach the depth of net you desire.

Thread suspension line casing along the bottom and sides of the net to strengthen it and to make it easier to handle.

You are now ready to set the net. But first, attach small pieces of wood (floats) to the top and stones (anchors) to the bottom of the net.

Secure the net to poles or saplings in the water near the bank so that the net is set at a 45° angle to the current (figure 7-9). This angle will help to keep floating refuse from catching in the net.

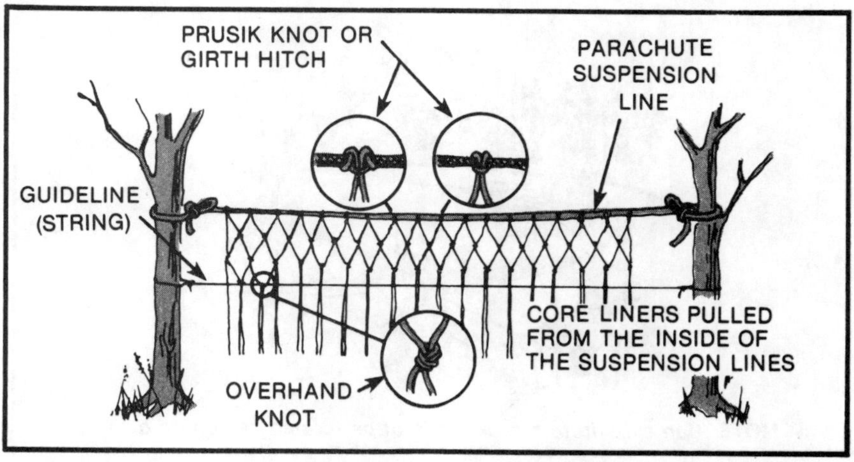

Figure 7-4. Making a gill net.

If you have another person with you, you can attach poles to the ends of the net. The two of you can then work the net up and down the stream. Be sure to pull up the net every few minutes and remove the fish before they can escape.

Use fish traps and baskets. These can be used for either freshwater or saltwater fish. They take a lot of time and effort to make, however, and are difficult to carry when you move on. Shown in figures 7-6 and 7-7 are some types of fish traps you can make.

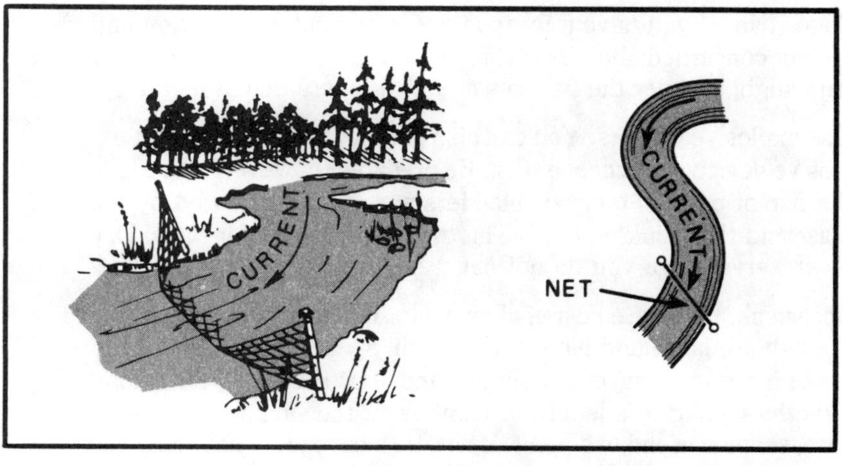

Figure 7-5. Setting a gill net in the stream.

NOTE: *Can substitute a sack made of perforated parachute gores.*

Figure 7-6. Setting a fish trap in a stream.

When setting traps for freshwater fish, keep in mind the habits listed on page 7-5.

You can also use traps to catch saltwater fish as schools regularly approach the shore with the incoming tide and often move parallel to the shore. Pick a trap location at high tide and then build your trap when the tide is low. On rocky shores, use natural rock pools. On coral islands, use natural pools on the surface of reefs by blocking the openings as the tide recedes. On sandy shores, use sandbars and the ditches they enclose. Build the trap as a low stone wall extending outward into the water and forming an angle with the shore.

Shoot fish. If you have a weapon and sufficient ammunition and are not concerned about revealing your location, try shooting fish. Aim slightly under the fish in water that is less than 3 feet deep.

Use explosive devices. You can also get fish by tossing a light explosive device in a school of fish. Be prepared, however, to retrieve the fish at once as their air bladders are usually ruptured by the blast and they sink quickly. This will supply food for days. Dry or preserve those you do not eat fresh (page 7-40).

Spear fish. If you are near shallow water (about waist deep) where the fish are large and plentiful, you can spear them. It is easy to make a spear using materials at hand. For the shaft use a long straight sapling or a length of bamboo. If the sapling has a solid core, shape one end to a point. If not, tie a bayonet, a pointed piece of metal, a sharpened bone, a knife, or thorns to it. With bamboo, shape two points just below the joint.

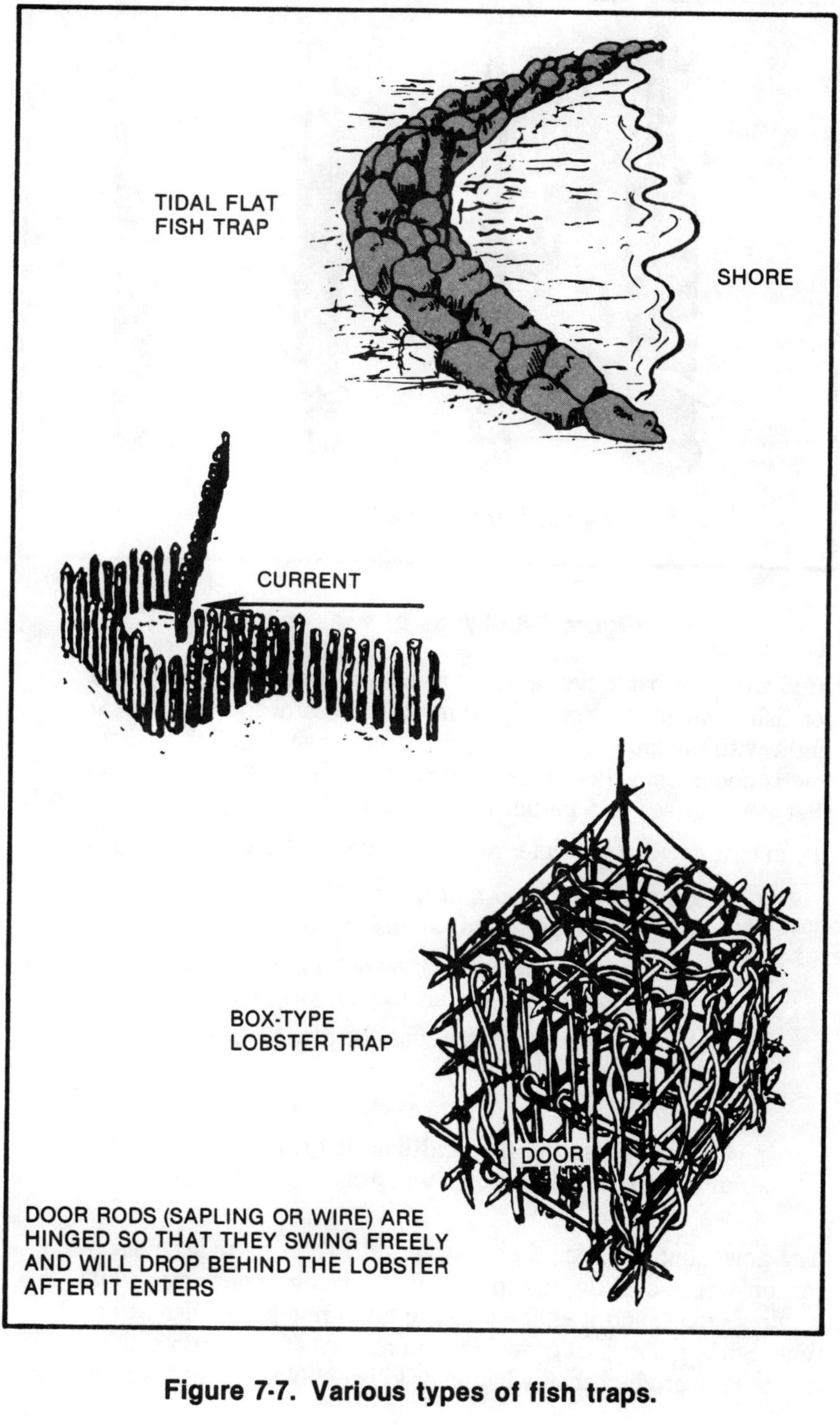

Figure 7-7. Various types of fish traps.

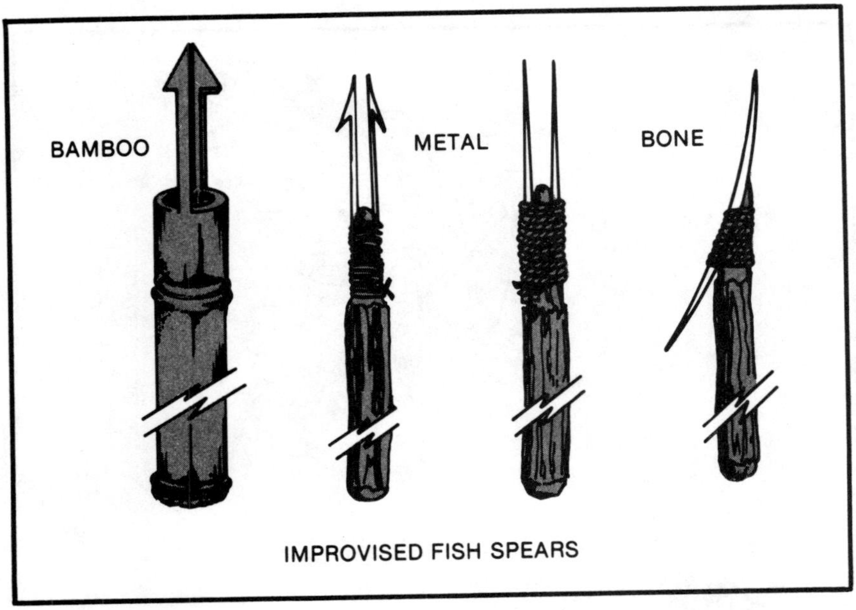

Figure 7-8. Types of spear points.

Find a rock or bank over a fish run and wait patiently and quietly for fish to swim by. You may be more successful spearing fish at night with the aid of a torch. Light attracts fish, highlights them, and reflects from their eyes. It also lights the stream bottom so that you can see and gather other aquatic life.

If you have to go in the water to a place where fish seem to gather—

- Wade very slowly to that place. All movement must be slow to cut down on vibrations in the water.
- Put the spear in the water and wait a few minutes for the fish to become accustomed to your presence.
- Move as close to your target as you can, keeping the spear in the water.
- Position the spear over and as close to the fish as you can.
- Quickly spear the fish and hold it firmly on the bottom until you can reach down and grasp it to lift it out of the water.

Try poisoning fish. Another way to get fish is by using poison. Poison works quickly, it allows you to remain undercover while it takes effect, and it enables you to get a number of fish at one time. Some plants that grow in warm regions of the world contain rotenone, a product that will stun or kill cold-blooded animals but

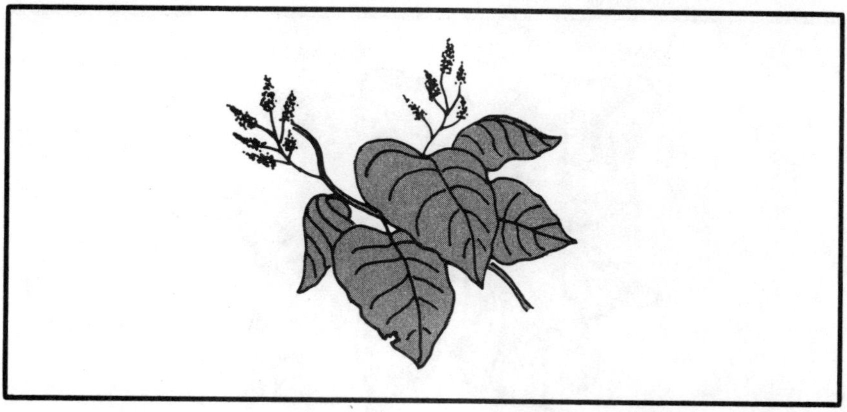

Figure 7-9. Anamirta cocculus.

will not affect persons who eat the animals. The best place to use rotenone, or rotenone-producing plants, is in ponds or at the headwaters of small streams containing fish. Rotenone acts quickly on fish in water 70⁰ F or above, and the fish rise helpless to the surface. It works slowly in water 50⁰ to 70⁰ F and is ineffective in water below 50⁰ F. The following plants, used as indicated, will stun or kill fish:

- Anamirta. This woody vine grows in southern Asia and on islands of the South Pacific. Crush the bean-shaped seeds and throw them in the water.
- Croton tiglium. This shrub or small tree grows in open waste areas on islands of the South Pacific. It bears seeds in three-angled capsules. Crush the seeds and throw them into the water.

Figure 7-10. Croton tiglium.

Figure 7-11. Barringtonia.

- Barringtonia. These large trees grow near the sea in Malaya and parts of Polynesia. They bear a fleshy one-seeded fruit. Crush the seeds and bark and throw them into the water.
- Derris. This large genus of tropical shrubs and woody vines is the main source of commercially produced rotenone. Powder or macerate the roots and mix with water. Throw a large quantity of the mixture into the water.

Figure 7-12. Derris eliptica.

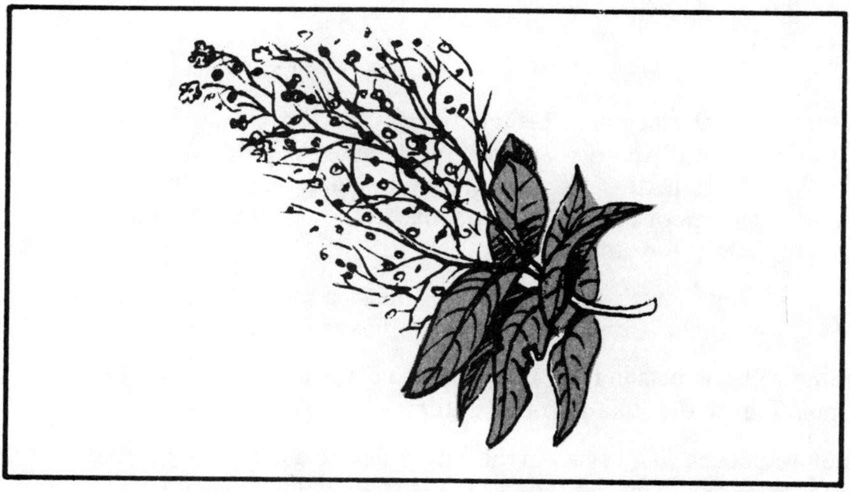

Figure 7-13. Duboisia.

- Duboisia. This shrub grows in Australia and bears white clusters of flowers and berrylike fruit. Crush the plants and throw them into the water.

- Tephrosia. This species of small shrubs, which bears bean-like pods, grows throughout the tropics. Crush or bruise bundles of leaves and stems and throw them into the water.

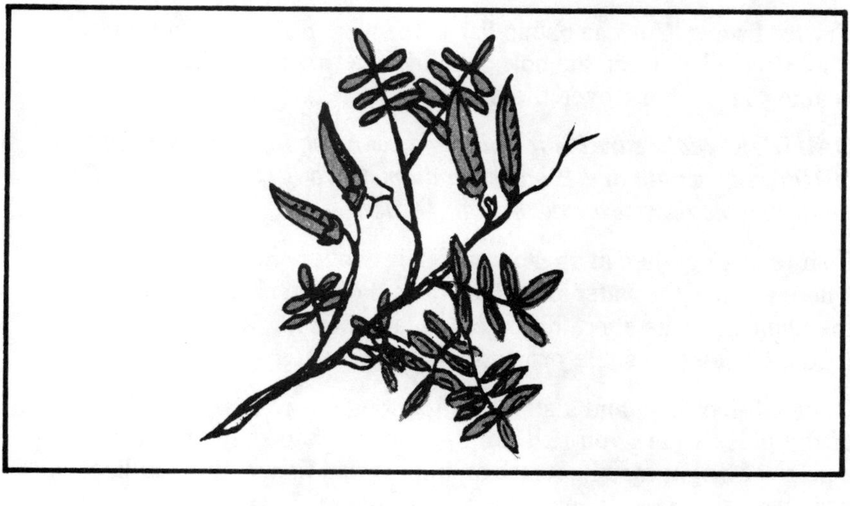

Figure 7-14. Tephrosia.

- Lonchocarpus. This genus of plants grows chiefly in tropical America and the West Indies as a tree or shrub. It has feather-like leaves and its bloom may be white or shades of red. Powder or macerate the roots and throw into the water.
- Green husks from hickory nuts, butternuts, or black walnuts. Crush and throw the husks into the water.

Lime will also poison fish. You can burn coral and seashells to get lime. Throw the lime into the water.

Before putting fish poison in the water, devise a means of picking up the stunned fish. Perhaps a dip net is all you will need. Or you may need to build a blockade downstream to catch the fish.

Try chop fishing. If you are on a beach and light security is not a factor, you can use the chop fishing technique to get fish at night when the tide is low. You will need a torch, a means of lighting the torch, and a machete.

With the lighted torch in one hand and the machete in the other—

- Wade into the water to a point where the water is about 2 feet deep. The light will attract the fish.
- Hit the fish with the back of the machete blade so they are stunned and then pick them up.

Try ice fishing. You can obtain fish in the winter by fishing through a hole in the ice. Keep the hole open by covering it with brush and heaping loose snow over the cover.

CAUTION: Make sure the ice will hold your weight. Carry a pole 8 to 10 feet in length and 2 inches in diameter to help you get out of the water should you break through the ice.

Fish tend to gather in shallow water in winter. So cut ice holes where you feel the water is medium deep. Possible places are where the shelf near the shore drops off to lake bottom, at the edge of reeds, or close to some projecting rock formation.

Take a 3-foot pole and a string long enough to reach the bottom of the place where you fish. Make a spoon-shaped spinner from a piece of bright metal. Attach an improvised fishhook to the line just above the hook. When fishing, move the rod in an up-and-down motion in such a way that the bright metal object vibrates.

Figure 7-15. Ice fishing hole.

Another method is to place a rig similar to the one in figure 7-15 at several holes. When the flag moves to an upright position, remove the fish and rebait the hook.

Eels. Eels are fish with a snakelike appearance found throughout the world in freshwater and saltwater. They are smooth skinned and swim underwater. (Snakes are scaled and usually swim on top.) Eels are excellent eating. You can catch them during the day in muddy water or at night using the same methods as for catching fish at night. Eels are easily speared at night under a torchlight. After catching an eel, strike a sharp blow to the head to stun it. Eels, like catfish, should be skinned before cooking.

Poisonous Fish. Some fish are poisonous to eat due to alkaloids in their flesh or to poisonous foods they have eaten. Cooking does not destroy these toxins. There is no firm guide for identifying poisonous fish. Some characteristics, however, are smooth skin, rough skin, bristles, or spines rather than true scales; some puff up or inflate when disturbed. Poisonous fish are seldom found in the open sea, but live around rocky or coral reefs and muddy or sandy shores. Examples are the puffer, porcupine, cow, and thorn fish (figure 7-16), which contain toxic substances in their flesh. If there are natives in the area, observe what fish they eat.

No poisonous fish are known to live along the shores of the North Pacific and Arctic Oceans. The eggs of the sculpin, however, are deadly poisonous. Do not eat fish eggs found in clusters or clumps on rocks, logs, or reefs.

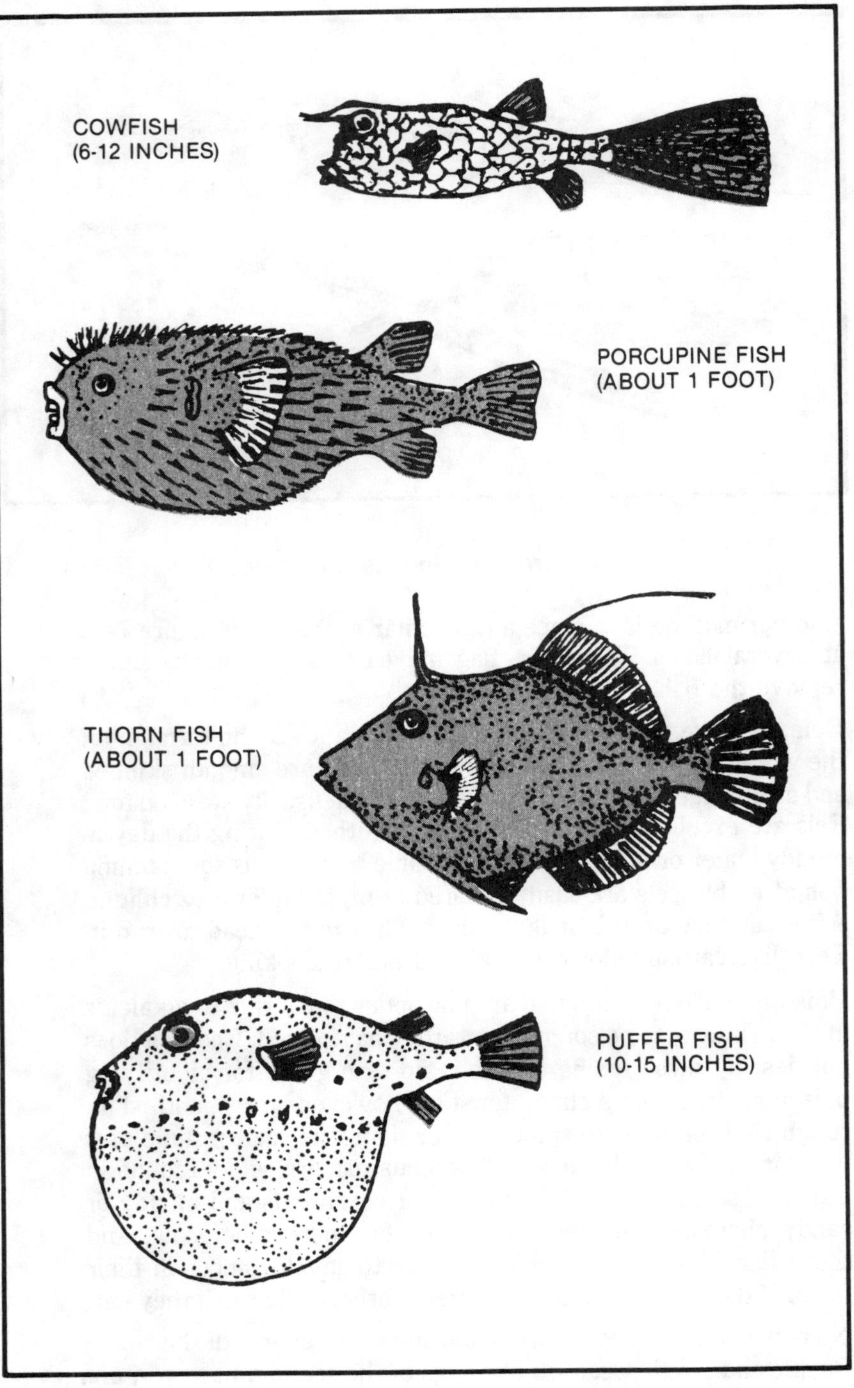

Figure 7-16. Fish with poisonous flesh.

Small Amphibious Animals

Frogs, newts, and salamanders inhabit areas surrounding freshwater in warm and temperate climates throughout the world.

Hunt frogs at night when you can locate them by their croaking. Club them or snag the larger ones on a hook and line. Skin a frog, cook and eat the entire body.

Newts and salamanders are found under rotten logs or under rocks in areas where frogs are abundant. Skin and remove the innards before cooking these amphibians.

Reptiles

Many reptiles are a good food source and are available in many parts of the world.

Snakes. All poisonous and nonpoisonous freshwater and land snakes are edible.

CAUTION: *Take extreme care in securing snakes as the bite of some poisonous snakes can be fatal. Even after a snake's head is cut off, its reflex action can cause it to bite, injecting poison.)*

The best times to capture snakes are in the early morning before temperatures get too high and in the late afternoon and early evenings before temperatures get too low. You can find snakes in any place where there is cover. Hit the snake with a rock, switch, or club to kill it or use a long stick to pin down its head to capture it. To pick up a snake, place your index finger on the top rear of its head with your thumb and middle finger on either side of the head behind the jaws. You must keep your finger on top of the snake's head to keep the snake from turning inside its skin and biting you.

CAUTION: *Take extreme care at all times.*

Lizards. These reptiles are most abundant in the tropics and subtropics, but they can be found most everywhere. The meat of all species is edible. The bite of two species, however, is poisonous. These two species—the gila monster and the beaded lizard—are found only in the American southwest, Central America, and Mexico. You can club lizards or snare them by attaching a grass or bark noose to the end of a stick. Skin large lizards and remove the entrails, then broil, roast, boil, or fry the meat. Small lizards may be impaled through the mouth on a stick and roasted without removing the entrails. The meat is done when the skin bubbles and cracks.

Turtles. Snapping turtles tend to hide or rest by crawling head first into holes dug into the bank. To locate a turtle in the bank, feel with your feet with your shoes on. When you locate the turtle, reach down and feel for the top of the shell. Grasp the tail, which is located under the serrated edge of shell, and pick up the turtle.

CAUTION: *Do not attempt to find turtles in areas where there are water moccasins or poisonous water snakes.*

CAUTION:*Take extreme care when attempting to catch snapping turtles. They can bite underwater as well as on land.*

Crocodiles and Alligators. These reptiles can be found around and in the lakes, rivers, and streams in tropical and warm temperate zones. The best way to kill them is by shooting. To skin them, heat them over a fire to loosen the plates. Broil or fry the meat.

Birds and Mammals

In some areas, birds and mammals may be more abundant than other forms of animal life. One important factor is *all are edible.*

Birds and mammals tend to congregate in their habitats. Places to look for them are—

- At the edges of woods and jungles.
- On trails, in glades, and in openings in a forest or jungle.
- On banks of streams and rivers.
- On lakeshores and seashores.

Birds and mammals are easily seen, and their eating, sleeping, drinking, and traveling activities are fairly regular. By observing their habits and activities, you can anticipate their movements. This will enable you to hunt or trap them successfully.

Birds. Birds lack a sense of smell, but they can see and hear exceptionally well. So to catch birds, you need to know some of their habits.

Nesting. The time when birds are least likely to flee from man is

Types of Birds	Frequent Nesting Places	Nesting Periods
Inland birds	Trees, woods, or fields	Spring and early summer in temperate and arctic regions. Year round in the tropics.
Cranes and herons	Mangrove swamps or high trees near water	Spring and early summer.
Some species of owls	High trees	Late December through March.
Ducks, geese, and swans	Tundra areas near ponds, rivers, or lakes	Spring and early summer in arctic regions.
Some sea birds	Sandbars or low sand islands	Spring and early summer in temperate and arctic regions.
Gulls, auks, murres, and cormorants	Steep rocky coasts	Spring and early summer in temperate and arctic regions.

Figure 7-17. Bird nesting places.

when they are nesting. Knowing where and when the various birds nest (figure 7-17) will make catching them easier.

During their nesting period, birds return often and regularly to their nests if the surroundings are not greatly disturbed. If you are patient, you can catch them during their nesting. You can actually touch some birds while they are incubating eggs or brooding young. You can catch some with a noosing wand (figure 7-18). Build a blind of vegetation. Remain still and quiet in the blind and wait for the

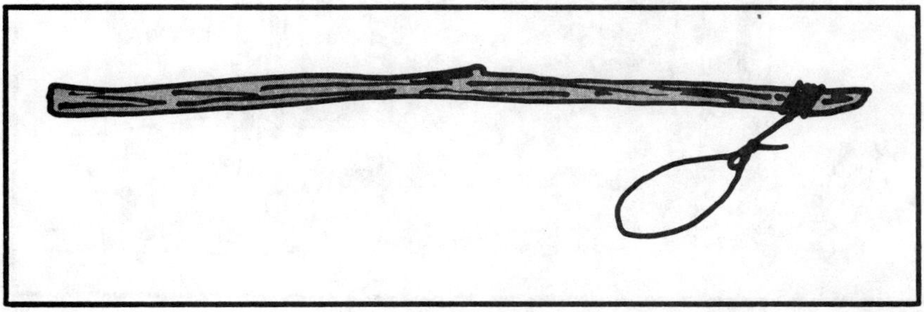

Figure 7-18. Noosing wand.

birds to return. Drop the noose over the bird's head, quickly pulling the noose up and back. Be prepared to club owls defending their nests. For birds that nest in hollow trees, block them in or noose them as they go in or out.

Do not overlook bird eggs. They are edible when fresh, even with embryos. Various species of ground-nesting birds, such as grouse, turkey, pheasant, and waterfowl, do not begin incubating their eggs until a full nest of eggs is laid. By carefully taking all but two or three eggs from the nest every few days, you will have a supply of fresh eggs for a week or two. Do not disturb the nest and do not remove all the eggs as the female will desert the nest. Do not kill the female during its nesting period if you want to keep your egg source.

Flying to Roost and Roosting. At dusk watch for birds going to roost. Some songbirds use the same flyway each day. Once you find their flyway, you can place a survival or mist net in it to catch some birds. Or you can locate the tree in which they roost, fix a net between two poles on one side of the tree, and then scare the birds from the other side of the tree so that they fly into and become entangled in the net.

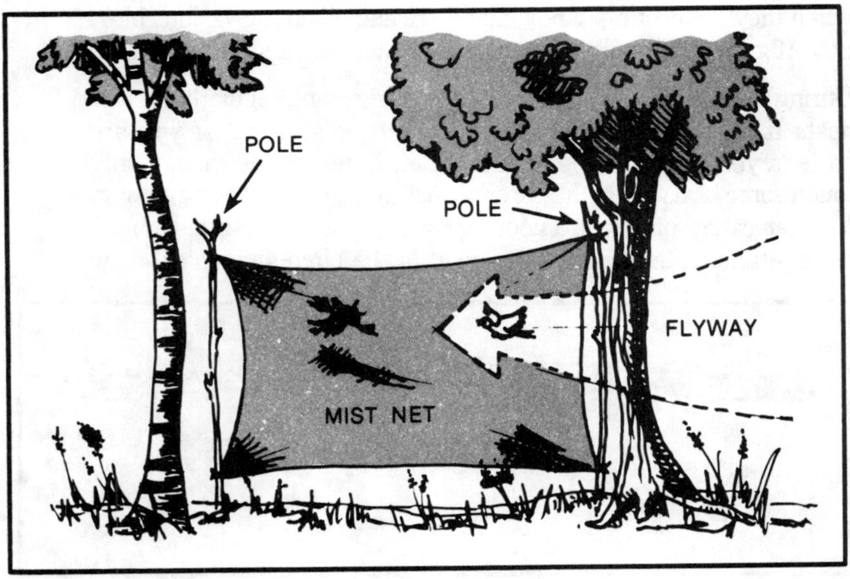

Figure 7-19.Catching birds in a net.

Larger birds, such as grouse, that roost in trees can be grabbed by the feet, noosed, or clubbed. Quail, grouse, and pheasant fly only a short distance when disturbed at roost, so if you fail to catch one on your first try, follow them and try again.

Moving. Waterfowl, quail, pheasant, turkeys, pigeons, and gulls tend to gather and move in flocks. Use a bola or an 18- to 24-inch throw stick to hit birds in a flock.

Feeding. On the beach you can sometimes catch shorebirds, herons, and fish-eating ducks by baiting a hook with a fish and placing the fish on the beach or trolling it in the water. A gorge hook (figure 7-1) placed inside the fish works well. Sometimes you can catch birds on the beach using a leg noose or snare.

Inland birds are easily caught using an Ojibwa bird snare (figure 7-20). To build this snare—

- Select a small tree or large sapling in an area inhabited by birds. (Or you can drive a large limb securely in the ground in an area in which birds swarm.)
- Cut the top off the sapling so that a stump 3 to 4 feet high remains.
- Shape the top of the stump to a point so a bird cannot land on it.
- Make a small hole through the stump near the top. Make the hole square on the side where you plan to stick the perch.
- Cut a stick 7 to 8 inches long for the perch. Square off one end to fit loosely in the hole in the stump. Shape the other end to a point to secure the bait.
- Cut a cord (suspension line, for example) 3 to 4 feet long and make a noose at one end. The noose should be 6 to 7 inches in diameter.
- Tie a knot in the cord just beyond the noose. (This knot is to hold the perch in position until a bird lands on the perch.)
- Pull the running end of the cord through the hole in the stump until the knot reaches the hole. Insert the baited perch in the hole and use the knot in the cord to hold the perch lightly.
- Drape the noose over the perch.

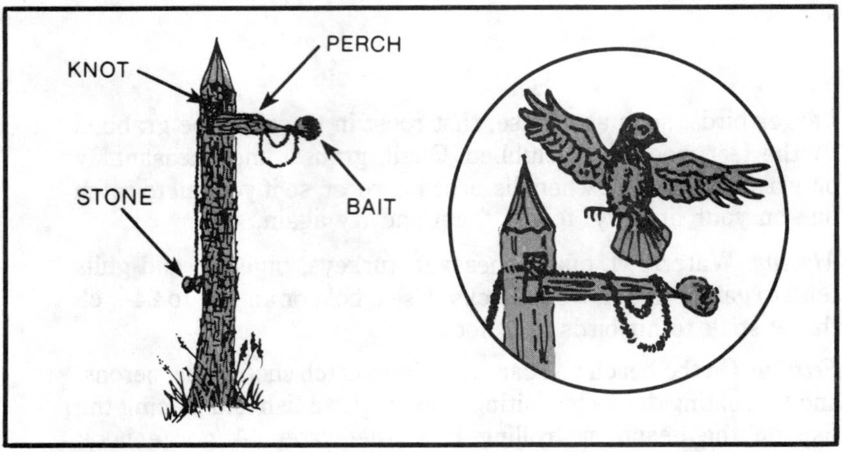

Figure 7-20. Ojibwa bird snare.

- Tie a stone or small log to the other end of the cord. The stone should be heavy enough to pull the noose when a bird lands on and dislodges the perch, but not so heavy that it jerks the noose tight. This might break off the bird's legs, allowing it to fly away.

You can also catch birds using perch snares (figure 7-21). Place several light perch snares on a tree limb or any horizontal bar. Arrange the perch snares so they overlap. Place bait on the limb or horizontal bar so that when the bird lands, its feet will become ensnared. Oftentimes one caught bird will attract others. This works well with blackbirds.

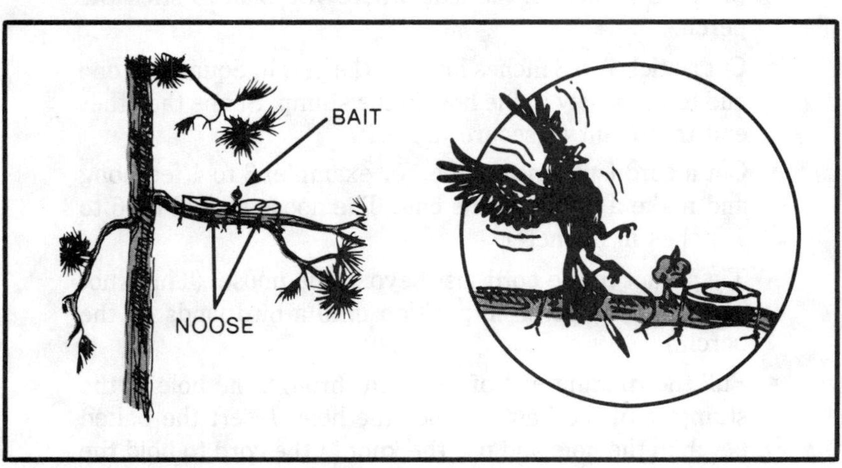

Figure 7-21. Bird perch snares.

Mammals. Game animals detect danger through their senses of sight, hearing, and smell. If you are unfamiliar with the game in the area, you should assume that the animals are naturally wary and their senses of sight, hearing, and smell are keen.

Many mammals travel on trails and runways. By observing their tracks, feces, runways, trails, dens, and feeding marks, you can tell the types of animals in the area and their relative abundance. This knowledge will help you in hunting or trapping them.

To be sure you get meat to eat, you must know how to *find, approach,* and *shoot* and *trap* game.

Finding and Approaching Game. Look for fresh signs, such as tracks, beds, and warm or moist droppings, to indicate the recent presence of game. *Keep in mind* that seeing your quarry before it sees you is one of the prime factors in hunting.

Look cautiously over ridges, whether in the woods or in the open, checking first the distant and then the closer ground.

Locate a water hole, feeding ground, or well traveled trail and wait quietly for the game to come to you.

Move up or across wind, never downwind. This applies equally to looking for and stalking game in either open or forested areas.

Move as silently as possible in densely forested areas that limit your range of vision. Any noise you make, for example, treading on dry sticks and leaves, will alert animals to your presence.

Keep undercover in areas where game can easily see you. If your quarry sights you and does not flee, stop until it resumes feeding. Then zigzag back and forth across your line of approach. Stop when the animal looks up.

Get above mountain game if possible. They are less likely to suspect danger from above.

Shooting Game. The vulnerable spots on many animals are the head, neck, or just back of the shoulder. Aim for one of these. Take your time to make the shot your very best. If you wound an animal and it runs, slowly follow its blood trail. A severely wounded animal will normally lie down soon if it is not followed. Give it time to weaken so it cannot rise. Approach slowly and complete your kill.

Catching Small Animals. You can sometimes get the following animals without the aid of a gun or a snare:

- Armadillos—Run down, catch by hand, or club.
- Rabbits/hares—Club or hit with a throw stick, rock, or bola. Hares in northern areas when jumped may run only 4 to 5 feet and stop.
- Mice/lemmings/moles—Step on or club. These animals are often found under rocks or logs in summer and in snow banks in winter.
- Squirrels—Use a squirrel pole (figure 7-22).
- Skunks—As the skunk is feeding or traveling, get within its line of track so that it will pass within your reach. Approach it carefully, grab its tail, and quickly lift it off the ground. It cannot spray when its feet are off the ground, and it seems unable to bend to bite when it is held by the tip of its tail. Club it. If possible, submerge it in water to cool before cleaning it, so that the glands will get hard. This will make them easier to remove and cause less odor emission.
- Opossums—Catch by tail; club when possible.
- Groundhogs/woodchucks/marmots—Place a doubled-wire snare (figure 7-23) at the animal's hole so that you can snare it entering or leaving its hole. Or you can dig the animal out of its hole and hit it with a club or rock. In digging out, when you get close to the animal, you can sometimes insert a belt with buckle or loop, which the animal will bite and hold onto. You can then pull it out. If you find the animal in the open, run it down; when it turns to fight, club or drop-kick it.
- Porcupine—When it is on the ground, club it. When it is in a tree, knock it down, hit it with a rock, or pin it down with a forked stick and pick it up by the long hairs at the end of the tail. Porcupines cannot throw quils. ***CAUTION:*** *Quills are barbed and will stick on contact.*
- Beaver—Wait for the beaver to come on land, then club it, drop-kick it, hit it with a rock, or catch it by the tail.

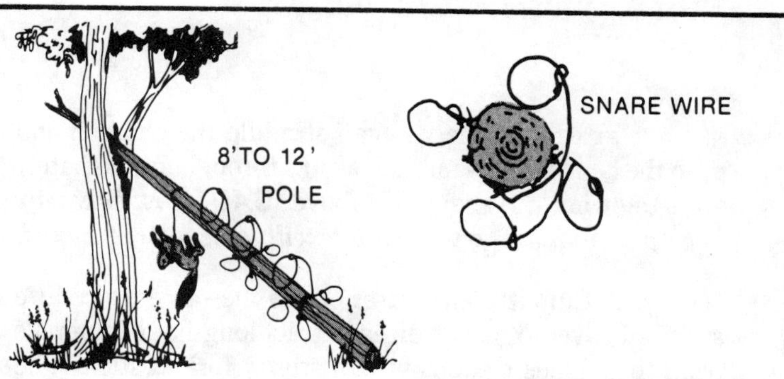

1. Lean an 8- to 12-foot pole against a tree that squirrels use.
2. Using snare wire, make three or four nooses. Attach the nooses to the pole.

Figure 7-22. Squirrel pole.

It is a sturdy animal, so if you catch it by the tail, swing it in a pendulum motion until it begins to relax, then swing it against a tree or the ground or use a noose to kill it. Another way to get a beaver is to dig out the beaver dam so the water drains. The beaver will come to inspect the

Figure 7-23. Doubled-wire snare.

damage at which time you can straddle the channel and grab the beaver by its tail as it swims through. Immediately start to swing the beaver as above. **CAUTION:** *Take care to keep it from biting you. Its bite will leave a large wound.*

- Muskrat—Club it in the water or use the same procedure as for a beaver. You can often extract long-haired animals from their holes or dens with a sturdy forked stick. The tines of the fork should be 2 to 3 inches apart, 1-1/2 to 2-1/2 inches long, and 1/2 to 1 inch in diameter and sharpened. Jam the forked stick into the animal, twisting tightly while holding pressure. The stick will hold the hide so you can pull out the animal. Be ready to club it immediately.
- Mongoose—Hit with a club or rock, or catch it with a hand-activated snare. The animal's lightweight may make it difficult to trip the snare. **CAUTION:** *These animals are quite ferocious and can inflict a nasty bite.*

You can also let predators hunt for you but you must be observant. Watch for both flying and foot predators hunting, stalking, or killing game. Oftentimes you can hear smaller animals being killed. By rushing the predator, you may cause it to drop its prey, which you can recover.

If a large predator, such as a bear or large cat, has been scared away from a fresh kill, building a fire overnight should keep the irate predator away. Most predators will avoid fires and humans.

Trapping Game. If you are without a gun, or if you are in an area where you must keep quiet, you can build snares, traps, or deadfalls to catch game. But first, you must decide from the animal signs you have seen what animal to catch and then how to catch it. Successful trapping, same as successful hunting, depends on your knowing the habits of the animals. Many rodents and carnivores, for instance, are active only at night; hoofed animals forage both day and night.

The type of game you decide to trap and the habits of that game determine the type of snare or trap you should use. But regardless of the type you use, you should cover up your scent. You can do this by soaking the snare material in a stream, rubbing the material with cold ashes, or spreading urine from an animal's bladder on the material to lure the same type of animal into the trap. Weathering (time and/or rain) or smoke will also cover your scent.

Although salt is not a scent killer, it is an effective lure in areas away from the coast.

One means of capturing small game is with a snare. A snare is a noose that will slip and strangle or hold any animal caught in it.

To make the noose, use material such as the inner core strands of parachute shroud lines, hide strips from previously caught animals, or the bark of small hardwood saplings. Or you can use wire; booby trap wire works fine.

You can use a noosing wand (page 7-22) to snare an animal as it comes out of its burrow. Use the same procedure as for snaring a bird.

The drag noose snare, however, is usually more desirable in that it allows you to move away from the site. It is also one of the easiest to make and fastest to set. You can set it, then merely check it periodically. It is especially suitable for catching rabbits. All you have to do is find a game trail that indicates recent use. (You may find it is easier to detect small game trails by lying on your stomach and observing the terrain at ground level.) Find a spot along the trail where there are bushes or brush on both sides of the trail. Then make and set your snare. To make the snare (figure 7-24) use a string or wire and a sturdy branch as follows:

- Make a loop in the string using a bowline knot . (When using wire, secure the loop by intertwining the end of the wire with the wire at the top of the loop.)
- Pull the other end of the string (or wire) through the loop to form a noose that is large enough for the animal's head but too small for its body.
- Tie the string (or attach the wire) to a sturdy branch. The branch should be long enough to span the trail and rest on the brush or other support you have selected.

Now set the drag noose. If there is no suitable spot with brush on which to hang the drag branch, use two short forked sticks as a stand. To hold the string noose open, cut small notches in the forked sticks, or stick two small twigs in the ground.

A snared animal will dislodge the drag stick, pulling the stick along until it becomes entangled in brush. The animal's attempt to escape will tighten the noose, strangling or at least holding the animal.

There are several versions of this type snare. One version is to—

- Place a sturdy stick or small log across the trail, supporting one end with a forked stick.
- Tie a noose to the small log so that it hangs over the trail. (Again, adjust the size of the noose according to the size of the animal you want to catch.)
- Build a fence or place obstacles on either side of the noose to channel the animal into it. *Do not use green twigs or branches* as the animal may stop to eat them and bypass the noose. This "channeling" or "funneling" should look as natural as possible to blend with the environment and should extend approximately 6 inches on either side of the snare.

Another version is to build a fence on opposite sides of the trail, cutting notches on the sides of the two inner sticks. Spread open the noose, and place the sides in the notches on the sticks. You can balance the drag stick on top of the fence or lay it to the side.

Figure 7-24. Drag noose.

The snare loop, locking type, is a noose (figure 7-25) that will lock when pulled tight, ensuring that the snared animal cannot escape. To make it, use lightweight wire, such as wire from a vehicle or aircraft electrical system or any light, woven electrical wire. If this type wire is unavailable, you can use trip wire. To make this snare—

- Cut a piece of wire twice the length of the desired snare wire.
- Double the wire and attach the running ends to a securely placed object, such as the branch of a tree.
- Place a stick about 1/2 inch in diameter through the loop end of the wire.
- Hold the stick so the wire is taut and turn the stick in a winding motion so that the doubled wire is twisted together uniformly. You should have four to five twists per inch of twisted wire.
- Detach the running ends of the wire from the branch and remove the stick from the loop.
- Make a figure 8 in the 1/2-inch loop by twisting the loop over itself, then fold the figure 8 so the small loops are almost overlapping.
- Run the loose wire ends through these loops. This forms a stiff noose that is strong, yet pliable enough to close easily when pulled.
- Tie the loose ends of the wire to the stick or branch you are using to complete your snare.

This is an excellent snare for catching large animals or for using in extreme cold climates where the wire could become brittle.

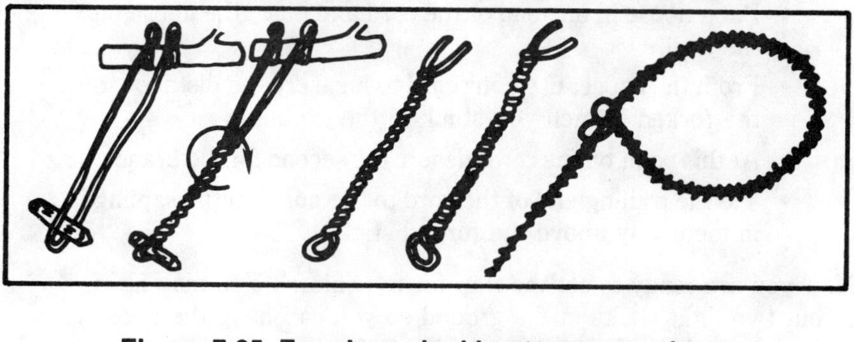

Figure 7-25. Forming a locking-type snare loop.

Remember, it's up to you to decide the type of snare to use based on—

- Your situation.
- The type of game prevalent in the area and the habits of that game.
- The type of material you have and the type of natural material available in the area.

Following are other types of snares that you might decide to use.

The hanging snare is another means of catching small game. It is one that lifts the caught animal off the ground so that the animal strangles. There are several different versions of the hanging snare, the differences being in the triggering devices used. One that is sometimes referred to as a fixed snare is easily made. It requires a sapling, a cord, and two small forked branches. To make it—

- Cut two small forked pieces of branch. Make sure that the forks will hook together. *NOTE: You can use two straight pieces and cut notches in them. This, however, takes more time.*
- Find a sapling near the animal trail you have selected.
- Pull the top of the sapling down toward the trail, but do not pull it so far that you destroy its tension.
- Mark the spot on the ground immediately below the point on the sapling where you plan to tie the cord. Do not tie the cord to the end of the sapling.
- Stick one small forked branch securely into the ground at the marked spot.
- Put a noose in one end of the cord and place the noose on the trail.
- From the noose, use your cord to measure the distance to the forked branch you stuck in the ground.
- At this point on the cord, tie it to the second forked branch.
- Tie the trailing end of the cord to the point on the sapling immediately above the forked stick.

Now you are ready to set the trap for the animal. You may have to put two light sticks in the ground so you can hang the noose if the brush along the trail is unsuitable for this purpose.

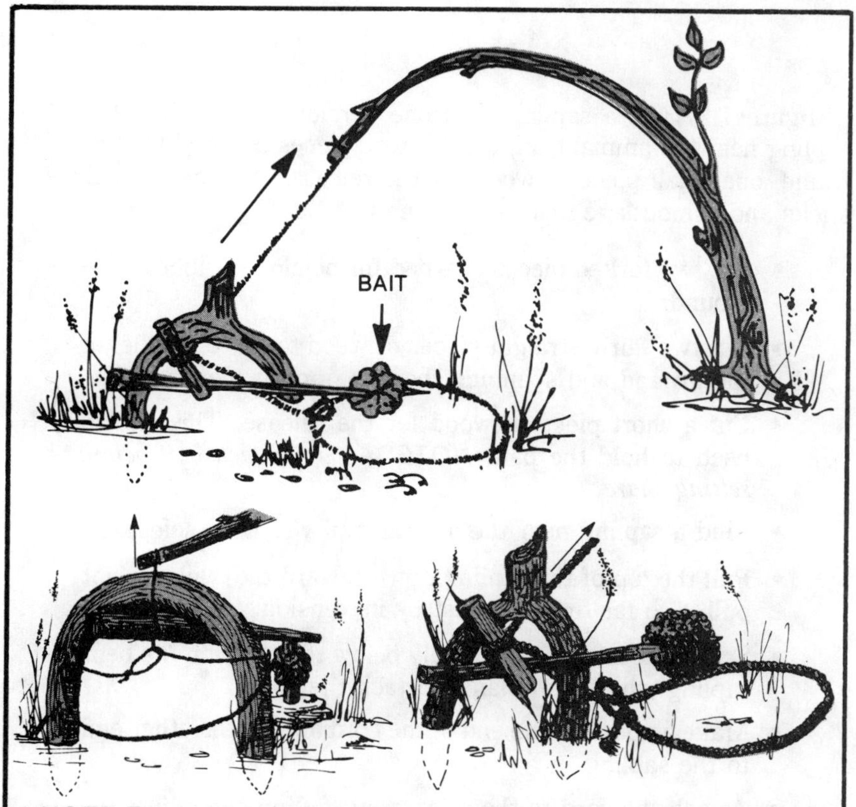

NOTE: Ensure that loop does not pass through arch of the forked stick.

Figure 7-26. Hanging snares using different triggering devices.

Another type triggering device for a hanging snare uses a forked stick and two straight sticks. One point of the fork must be long enough to stick securely in the ground. The other point need not, but can, touch or stick in the ground. One straight stick must be long enough to span and extend beyond the fork for attaching the bait. The shorter stick serves as the release. To make this snare—

- Tie the cord to the bent sapling and to the release stick.
- Bait the long stick.
- Place the long stick between the forked stick and the release stick, using the tension created by the bent sapling to hold it in place.

Another type of hanging snare is referred to as the treadle spring snare. Here again, there are different versions. The two versions in figure 7-27 use a sapling. For one version you will need a sapling near the animal trail, a cord, four pieces of wood for the frame, one small piece of wood for the release, and some small sticks and camouflage material for the treadle.

- Cut two forked pieces of wood for placing upright in the ground.
- Cut two fairly straight pieces of wood to place parallel to the ground and spanning the two upright pieces.
- Cut a short piece of wood for the release. This is also used to hold the bait. ***NOTE:*** *Always attach bait before setting snare.*
- Find a sapling near the animal trail you have selected.
- Pull the top of the sapling down toward the trail. Do not pull it so far that you destroy its tension.
- Set up the snare immediately below the point on the bent sapling where you plan to attach the cord.
- Make a noose in one end of the cord and tie the other end to the sapling.
- Attach the cord to the release stick. You can adjust the length of the cord by wrapping several times around the release stick.
- Set the snare as shown.
- On one side of the frame, rest one end of dry twigs or branches on the lower horizontal bar and the other end on the ground.
- Place other small twigs or leaves crosswise on top to reinforce and camouflage the platform.
- Place the noose on this platform.

You can make another version of the treadle spring snare by using a flexible branch. Bend the branch and stick each end in the ground as shown. If the soil is soft, make sure the ends of the branch stay securely in the ground. You can do so by placing heavy rocks at either end of the branch. Or you can take two long-forked branches and pound them securely in the ground over the ends of the bent branch.

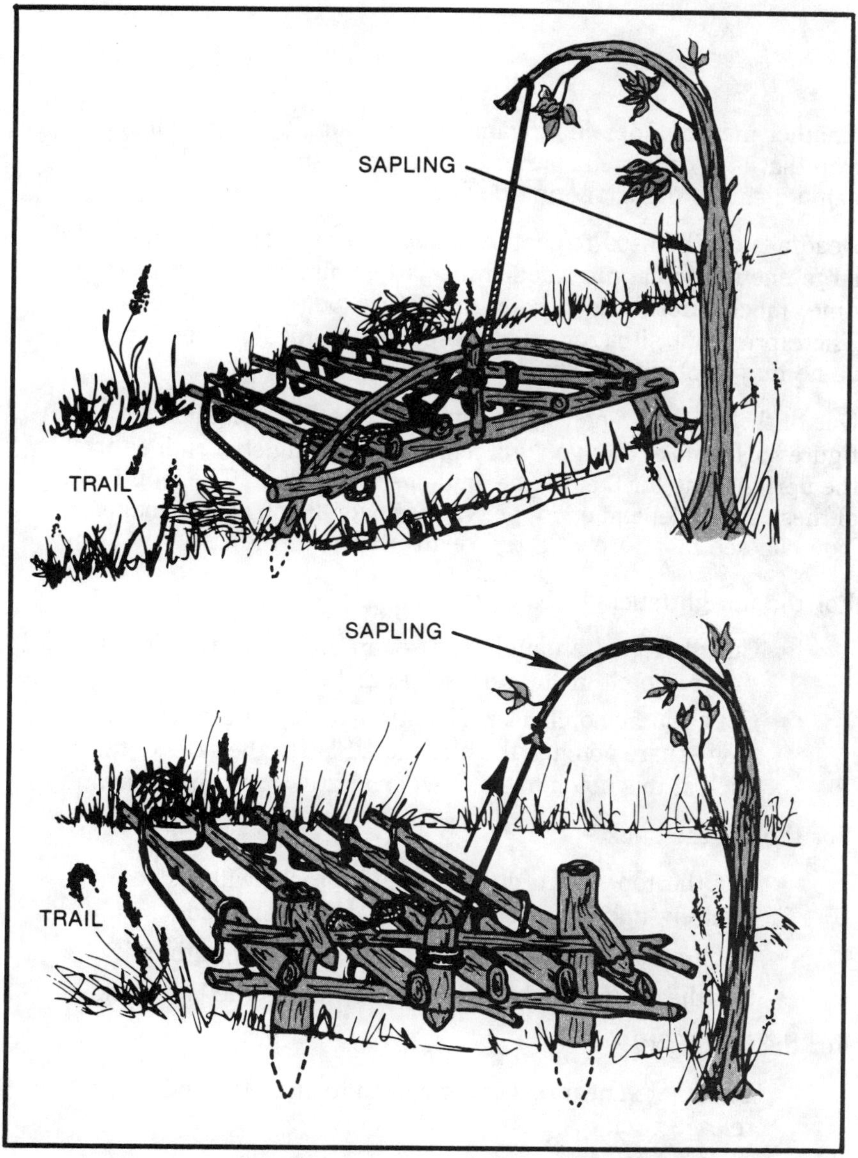

Figure 7-27. A treadle spring snare.

When setting snares, keep in mind that animals moving along trails normally have their noses slightly off the ground unless disturbed or startled. Therefore, you should set the snare so the bottom of the noose is no more than 2-1/2 to 3 inches above the ground for small animals, and between 24 to 36 inches above the ground for deer.

Another means of obtaining game is the deadfall. A deadfall is a trap that is constructed so that a weight falls on game when the game disturbs the triggering device.

Deadfalls can be used to capture many types of game, but those large enough to capture medium to large animals require much time, much effort, and usually more than one person to build. Therefore, your situation and the sparsity of medium and large game may prohibit a large deadfall.

One of the easiest triggering devices to make for a deadfall is the figure 4. To make this you will need three strong sticks: one for the upright, one for the release, and one for the bait. The lengths of these sticks depend upon the type of deadfall you plan to make. You can see in figure 7-28 the relative sizes of the sticks.

For the upright stick—

- Cut the top at an angle and square off tip so that it will fit into notch in the release stick.
- Cut a square notch near the bottom to fit into a correspon-ding square notch in the bait stick. Flatten the sides of the stick at this notch to ensure a good fit with the bait stick.

For the release stick—

- Cut the top so the deadfall (log or rock) will rest on it securely until the triggering device is disturbed.
- Cut a notch near the top in which to fit the upright stick.
- Cut the bottom end at an angle to fit into the bait stick.

For the bait stick—

- Cut a notch near one end in which to place the end of the release stick.
- Shape the other end to a point on which to secure the bait.
- Cut a square notch at the spot where it crosses the upright stick. The notches in the upright stick and the bait stick should fit firmly together.
- Get a stone or piece of wood on which to place the trig-gering device so that it will not sink into the ground from the deadfall weight.

You are now ready to set up your deadfall.

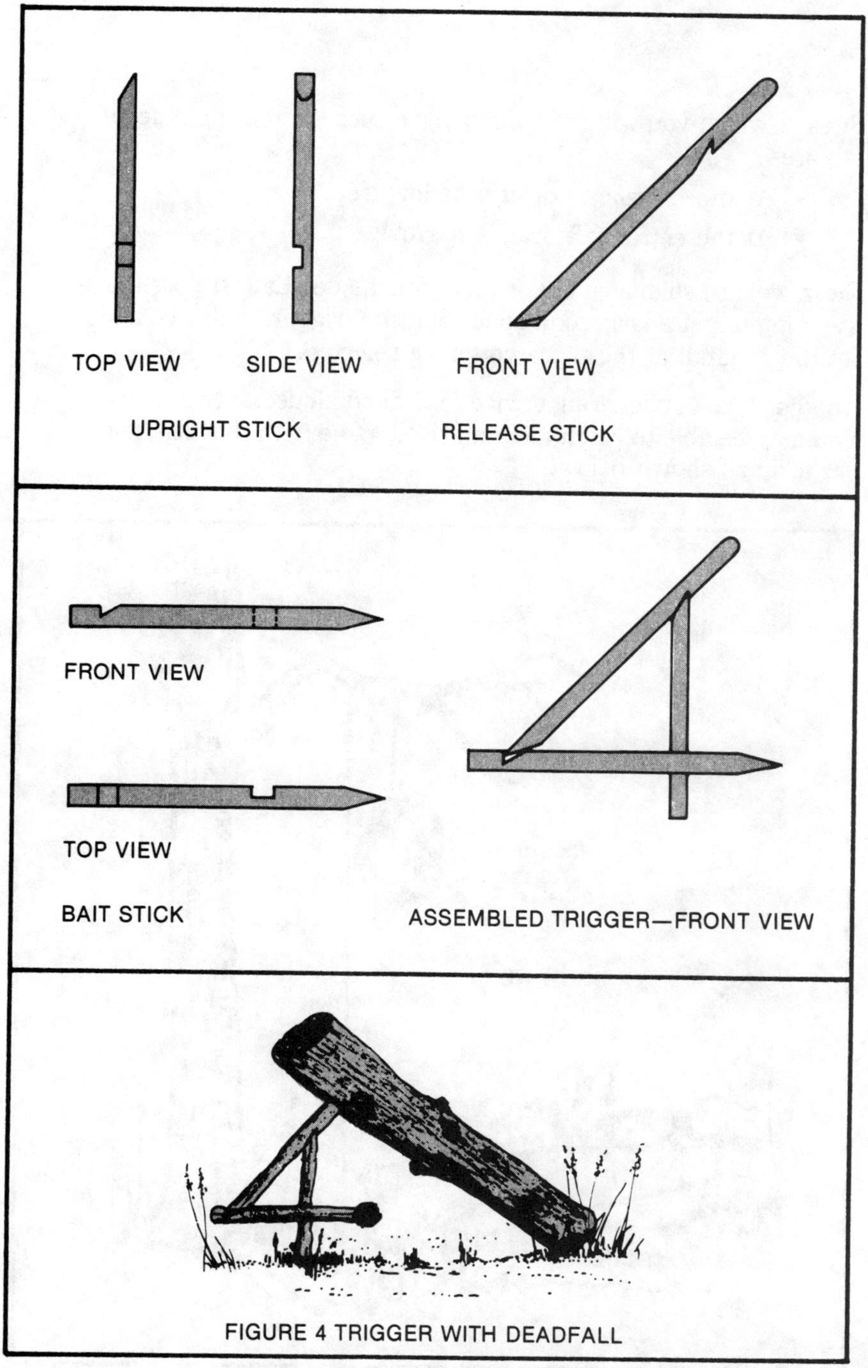

Figure 7-28. Figure 4 trigger.

Sites at which you will most likely have success in using a deadfall are—

- At the entrance/exit of a hollow tree, or
- At the entrance/exit of a burrow.

Get a weight, such as a log or rock, for the deadfall. If possible, use a log to get a larger kill zone. Bait the triggering device and set the deadfall at the place you have selected.

Another type of triggering device for a deadfall uses a trip string. You may be able to catch medium-sized game with this deadfall. Set it up as shown below.

Figure 7-29. Trip-string deadfall trap.

Preparing Fish and Game for Eating

You cannot afford not to know how to prepare fish and game for cooking and/or storing when you are in a survival situation. Improperly cleaning or storing can result in inedible fish and game.

Fish. On pages 7-18 and 7-19 you learned about poisonous fish that you must avoid. But just as important, you must know how to tell if fish are free of bacterial decomposition that makes the fish dangerous to eat. Although cooking may destroy the toxin from bacterial decomposition, do not eat fish that appear spoiled. Signs of spoilage are—

- A peculiar odor.
- A suspicious color. (Gills should be red to pink. Scales should be a pronounced—not faded—shade of gray.)
- A dent remaining after pressing the thumb against the flesh.
- A slimy rather than moist or wet body.
- A sharp or peppery taste.

Eating spoiled or poisoned fish may cause diarrhea, nausea, cramps, vomiting, itching, paralysis, or a metallic taste in the mouth. These symptoms appear suddenly 1 to 6 hours after eating. If you are near the sea, drink sea water immediately upon onset of such symptoms and force yourself to vomit.

Fish spoil quickly after death, especially on a hot day, so prepare fish for eating as soon as possible after you catch them.

Cut out the gills and large blood vessels that lie next to the backbone. (You can leave the head on if you plan to cook the fish on a spit.)

Gut fish that are more than 4 inches long. To do so, cut along the abdomen and scrape out the intestines.

Scale or skin the fish.

You can impale a whole fish on a stick and cook it over an open fire. However, boiling the fish with the skin on is the best way to get the most food value. The fats and oil are under the skin, and by boiling the fish, you can save the juices for broth. Any of the methods used for cooking plant food (page 6-6) can be used for cooking fish. Fish is done when the meat flakes off.

If you plan to keep the fish for eating later, smoke them using the same procedure as for smoking game (pages 7-45 and 7-46) or dry them in the sun. To prepare fish for smoking, cut off the heads and remove the backbones. Then spread the fish flat and skewer in that position. For skewers, you can use thin willow branches with the bark removed.

To dry fish in the sun, hang them from branches or spread them on hot rocks. When the meat has dried, splash it with sea water, if available, to salt the outside. Do not keep any seafood unless it is well dried or salted.

Snakes. To prepare a snake for eating—

- Grip the snake firmly behind the head and cut off the head with a knife.
- Slit the belly and remove the innards. (You can use the innards for baiting traps and snares.)
- Skin the snake. (You can use the skin for improvising, belts, straps, or similar items.)

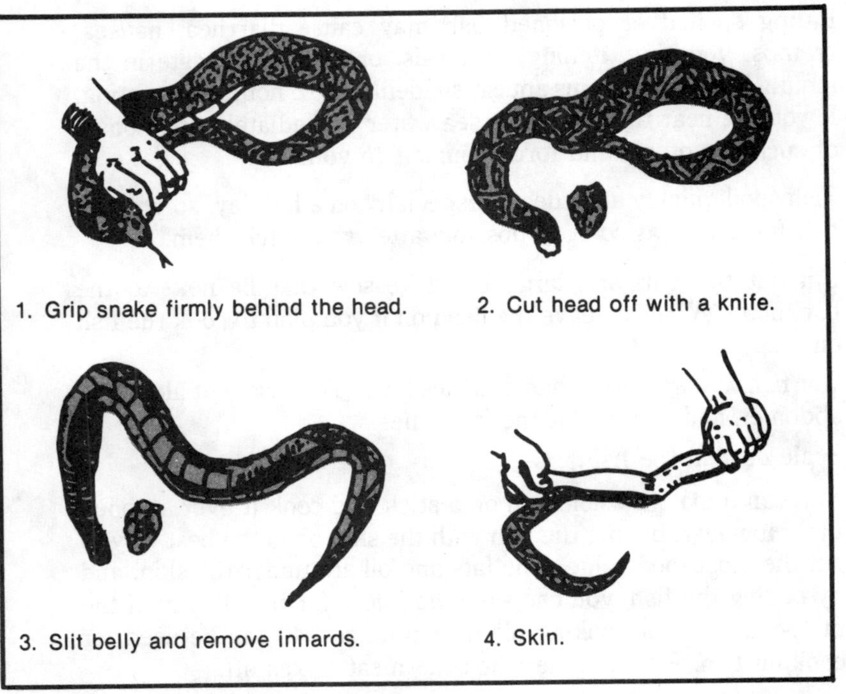

Figure 7-30. Cleaning a snake.

Fowl. Your first step after killing a fowl for eating or preserving is to pluck its feathers. If plucking is impractical, you can skin the fowl. Keep in mind, however, that a fowl cooked with the skin on retains more food value. Waterfowl are easier to pluck while dry, but other fowl are easier to pluck after scalding. After you pluck the fowl—

- Cut off its neck close to the body.
- Cut an incision in the abdominal cavity and clean out the insides. Save the neck, liver, and heart for stew. Thoroughly clean and dry the entrails to use for cordage.
- Wash out the abdominal cavity with fresh clean water.

You can boil fowl or cook it on a spit over a fire. You should boil scavenger birds such as vultures and buzzards for at least 20 minutes to kill any parasites.

Use the feathers from fowl for insulating your shoes, clothing, or bedding. You can also use feathers for fish lures.

Medium-sized Mammals. The game you trap or snare will generally be alive when you find it and therefore dangerous. Be careful when you approach a trapped animal. Use a spear or club to kill it so you can keep a safe distance from it.

After you kill an animal, immediately bleed it by cutting its throat. If you must drag the carcass any distance, do so before you cut off the hide so that the carcass is protected from dirt and debris that might contaminate it. Clean the animal near a stream if possible so that you can wash and cool the carcass and edible parts. Fleas and parasites will leave a cooled body so if the situation allows, wait until the animal cools before cleaning and dressing the carcass.

To skin and dress the animal—

- Place carcass, belly up, on a slope if available. You can use rocks or brush to support it.
- Remove genitals or udder by cutting circular area shown in figure 7-31.
- Remove musk glands at points A and B to avoid tainting meat.
- Split hide from tail to throat. Make the cut shallow so that you do not pierce the stomach.

- Insert your knife under the skin, taking care not to cut into the body cavity. Peel the hide back several inches on each side to keep hair out of the meat.

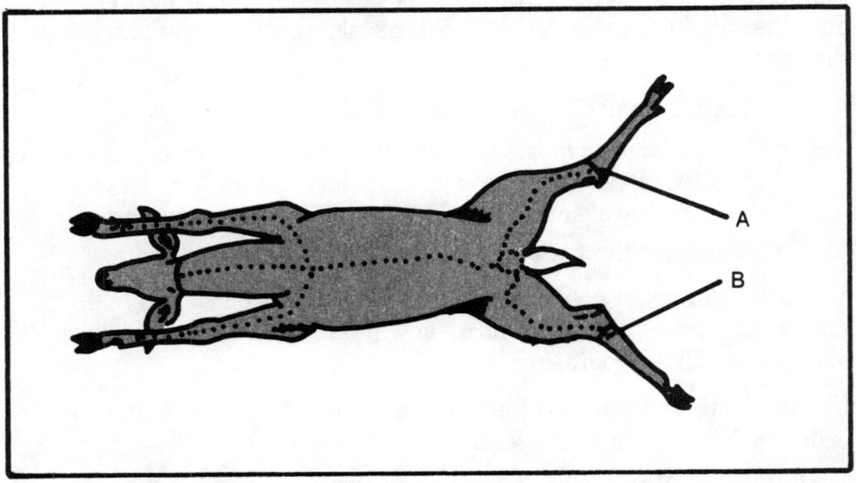

Figure 7-31. Skinning and butchering large game.

Figure 7-32. Skinning small game.

- Open the chest cavity by splitting the sternum. You can do this by cutting to one side of the sternum where the ribs join.
- Reach inside and cut the windpipe and gullet as close to the base of the skull as possible.
- With the forward end of the intestinal tract free, work your way to the rear, lifting out internal organs and intestines. Cut only where necessary to free them.
- Carefully cut the bladder away from the carcass so that you do not puncture the bladder (urine can contaminate meat). Pinch the urethra tightly and cut it beyond the point you are pinching.
- Remove the bladder.
- From the outside of the carcass, cut a circle around the anus.
- Pull the anus into the body cavity and out of the carcass.
- Lift or roll the carcass to drain all blood. ***NOTE:*** *Try to save as much blood as you can as it is a valuable source of food and salt. Boil the blood.*
- Remove the hide: Make cuts along the inside of the legs to just above the hoof or paw. Then peel the skin back, using your knife in a slicing motion to cut the membrane between the skin and meat. Continue this until the entire skin is removed.

Most of the entrails are usable. The heart, liver, and kidneys are edible. Cut open the heart and remove the blood from its chambers. Slice the kidneys and if enough water is available, soak or rinse them. In all animals except those of the deer family, the gall bladder—a small, dark-colored, clear-textured sac—is attached to the liver. Sometimes the sac looks like a blister on the liver. To remove the sac, hold the top portion of it and cut the liver around and behind the sac. If the gall bladder breaks and gall gets on the meat, wash it off immediately so the meat will not become tainted. Dispose of the gall.

Clean blood splattered on the meat will glaze over and help preserve the meat for a short time. However, if an animal is not bled properly, the blood will settle in the lowest part of its body and will spoil in a short time. Cut out any meat that becomes contaminated.

When temperatures are below 40 degrees, you can leave meat hanging for several days without danger of spoilage.

If maggots get on the meat, remove the maggots and cut out the discolored meat. The remaining meat is edible. Maggots, which are the larvae of insects, are also edible.

Blood, which contains salts and nutrients, is a good base for soups.

Although wild game has little fat, save all that you can and use it for making soap (page 3-3).

Thoroughly clean the intestines and use them for storing or smoking food or for lashings for general use. Make sure they are completely dry to preclude rotting.

The head of most animals contains a lot of meat, which is relatively easy to get. Skin the head, saving the skin for leather. Clean the mouth thoroughly and cut out the tongue. Remove the outer skin from the tongue after cooking. Cut or scrape the meat from the head. If you prefer, you can roast the head over an open fire before cutting off the meat. Eyes are edible. Cook them but discard the retina (this is a plastic like disc). The brain is also edible; in fact, some people consider it a delicacy. The brain is also used to tan leather, the theory being that the brain of an animal is adequate to tan its hide.

Use the tendons and ligaments of the body of large animals for lashings (page 4-7).

The marrow in bones is a rich food source. Crack the bones and scrap out the marrow. See Chapter 4 for using bones to make weapons.

If the situation and time allow, you should preserve the extra meat for later use. If the air is cold enough, you can freeze the meat. In warmer climates, however, you will need to use a drying or smoking process to preserve it. One night of heavy smoking will make meat edible for about 1 week. Two nights will make it remain edible for 2 to 4 weeks.

To prepare meat for drying or smoking, cut it with the grain in one-quarter inch strips.

To air dry the meat, hang it in the wind and hot sun out of the reach of animals; cover it so that blowflies cannot land on it.

To smoke meat, you will need an enclosed area—for instance, a teepee (figure 7-33) or a pit. You will also need wood from deciduous trees, preferably green. Do not use conifer trees such as pines, firs, spruces, or cedars as the smoke from these trees give the meat a disagreeable taste.

When using the parateepee or other enclosed area with a vent at the top, set the fire in the center and let it burn down to coals, then stoke it with green wood. Place the strips of meat on a grate or hang them from the top of the enclosure so that they are about 2 feet above the smoking coals.

To use the pit method of smoking meat, dig a hole about 3 feet (1 meter) deep and 1½ feet (½ meter) in diameter. Make a fire at the bottom of the hole. After it starts burning well, add chipped green wood or small branches of green wood to make it smoke. Place a wooden grate about 1½ feet (½ meter) above the fire and lay the strips of meat on the grate. Cover the pit with poles, boughs, leaves, or other material.

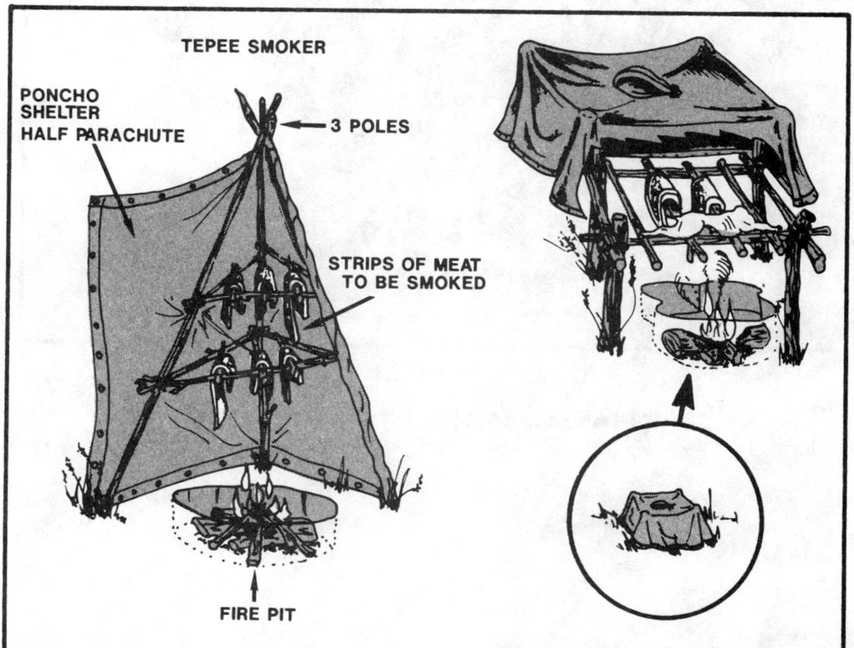

Figure 7-33. Smoking meat.

Properly smoked meat will look like a dark, brittle, curled stick. It is tasty and nutritious.

If mold forms on the meat, brush or wash it off before eating the meat. In damp weather, you will have to redry smoked or air-dried meat to prevent molding.

You can also use saltwater to help preserve meat. Soak the meat in saltwater, then dry it in the sun or over fire. The meat can spoil if too little salt is used or if it is dried improperly.

Figure 7-34. Smoking meat over a pit.

Chapter 8

Shelters

A shelter can protect you from the sun, insects, wind, rain, snow, hot or cold temperatures, and enemy observation. It can give you a feeling of well being; it can help you maintain your will to survive.

In some areas your need for shelter may take precedence over your need for food, possibly even your need for water. For example, prolonged exposure to cold can cause excessive fatigue and weakness (exhaustion), and an exhausted person may develop a "passive" outlook. To get rid of a passive outlook caused by prolonged exposure to cold, you must have adequate shelter. Therefore, shelter takes priority over other needs.

Shelter Site

When you are in a strange environment, start looking for a shelter site at least 2 hours before sunset. As you do so, keep in mind what you will need at the site. Two requisites are:

- It must contain material to make the type shelter you need, and
- It must be large enough and level enough for you to lie down comfortably.

When you consider these requisites, however, *you cannot ignore your tactical situation or your safety.* You must also consider whether the site—

- Provides concealment from enemy observation.
- Has camouflaged routes of escape.
- Is suitable for signaling if necessary.
- Affords protection against wild animals, rockfalls, and dead trees that might fall.

- Is free from insects and other pests.

You must also keep in mind the problems that could arise in your environment. For instance—

- In foothills, avoid flash flood areas.
- In mountainous terrain, avoid avalanche areas.
- Near bodies of water, avoid a site that is below the high water mark.

In some areas the season of the year has a strong bearing on the site you select. Ideal sites for a shelter differ in winter and summer. During cold winter months you will want a site that will protect you from the cold and wind but will have a source of fuel and water. During summer months in the same area, you will want a source of water, but you will want the site to be relatively free of insects.

Types of Shelters

As mentioned above, when you look for a shelter site, you have to keep in mind the type of shelter (protection) you need. But you also have to consider other factors:

- How much time and effort are needed to build the shelter?
- Will the shelter adequately protect you from the elements (rain, snow, wind, sun, etc.)?
- Do you have the tools to build it? If not, can you improvise tools from materials in the area?
- Do you have the type and amount of man-made materials needed to build it? If not, are there sufficient natural materials in the area?

To answer these questions, you need to know *how* to make various types of shelters and *what* material you need to make them.

Poncho Lean-to. It takes only a short time and minimal equipment to build this lean-to (figure 8-1). You need a poncho, 6 to 10 feet of rope, three stakes about 6 inches long, and two trees (or two poles) 7 to 9 feet apart.

Before you select the trees you will use (or decide where to place the poles), check the wind direction. Make sure the back of your lean-to will be into the wind.

To make the lean-to—

- Tie off the hood of the poncho. To do this, pull the drawcord tight; roll the hood longways, fold it into thirds, and tie it with the drawcord.
- Cut the rope in half; on one long side of the poncho, tie half of the rope to one corner grommet and the other half to the other corner grommet.
- Attach a dripstick (about a 4-inch stick) to each rope 1/4 to 3/4 inch away from the grommet. (These dripsticks will keep rainwater from running down the ropes into the lean-to. Using driplines is another way to prevent dripping inside the shelter. Tie lines or string about 4 inches long to each grommet along the top edge of the shelter. This allows water to run to and down the line without dripping into the shelter.)
- Tie the ropes about waist high on the trees (uprights). Use a round turn and two half hitches with a quick-release knot.
- Spread the poncho into the wind and anchor it to the ground. To do this, put three sharpened sticks through the grommets and into the ground.

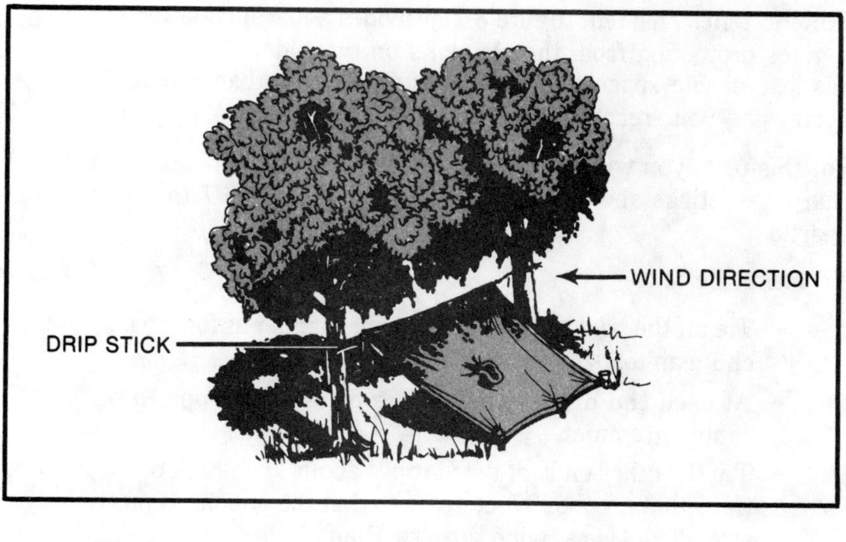

Figure 8-1. Poncho used for lean-to.

If you plan to use the lean-to for more than one night, or if you expect rain, make a center support for the lean-to. You can do this by stretching a rope between two upright poles or trees that are in line with the center of the poncho. Tie another rope to the poncho hood, pull it upward so that it lifts the center of the poncho, and tie it firmly to the rope stretched between the two uprights.

Another method is to cut a stick to place upright under the center of the lean-to. This method, however, will restrict your space and movements in the shelter.

To give additional protection from wind and rain, place boughs, brush, your rucksack, or other equipment at the sides of the lean-to.

To reduce heat loss to the ground, place some type of insulating material, such as leaves or pine needles, inside your lean-to.

NOTE: When at rest, as much as 80 percent of your body heat can be lost to the ground.

To increase your security from enemy observation, lower the silhouette of the lean-to by making two modifications: Secure the support lines to the trees knee-high rather than waist-high, use two knee-high sticks in the two center grommets (sides of lean-to), and angle the poncho to the ground, securing it with sharpened sticks as above.

Poncho tent. This tent (figure 8-2) provides a low silhouette, and it gives protection from the elements on two sides. However, it has less usable space and less observation area than a lean-to, decreasing your reaction time to enemy detection.

For this tent, you will need a poncho, two 5- to 8-foot ropes, six sharpened sticks about 6 inches long, and two trees 7 to 9 feet apart.

To make the tent—

- Tie off the poncho hood in the same manner as for the poncho lean-to.
- At each end of the poncho tie a 5- to 8-foot rope to the center grommet.
- Tie the other ends of these ropes about knee-high on two trees that are 7 to 9 feet apart so that the poncho is tautly stretched. Use a round turn, two half hitches, and a quick-release knot.

- Draw one side of the poncho tight and secure it to the ground with the three sharpened sticks stuck through the grommets.
- Follow the same procedure on the other side.

If you need a center support because of rain, attach one end of a line to the poncho hood and the other end to an overhanging branch so that the line is taut. Another type of center support is an A-frame set outside but over the center of the tent (figure 8-3). Use two 3-foot to 4-foot long sticks, one with a fork at one end to form the A-frame. Tie the drawstring on the hood to the A-frame so that the center of the tent is supported.

Figure 8-2. Poncho tent made using overhanging branch.

Figure 8-3. Poncho tent with A-frame support.

Shelters From Parachutes. If you have a parachute and three poles (and your tactical situation allows), you can make a *parachute tepee.* This type shelter is easy to make, takes very little time to make, provides protection from the elements, can act as a signaling device by enhancing a small amount of light from a fire or candle, and is large enough to hold several people and their equipment and to allow sleeping, cooking, and storing firewood.

You can make this tepee (figure 8-4) using parts of or a whole canopy of a personnel main or reserve parachute. If you are using a standard personnel parachute, you will need three poles 10 to 15 feet long and about 2 inches in diameter.

To make the tepee—

- Lay the poles on the ground and lash them together at one end.
- Stand the framework up and spread the poles apart to form a tripod.
- If you need more support, cut additional poles and stand them upright against the tripod. You need not lash them.
- Fold the parachute canopy in half. (If you are making a large tepee, cut the canopy in half from the *lower* lateral band to the *upper* lateral band.)
- Determine the wind direction and plan the location of your entrance so that it will be 90^0 or more from the mean wind direction.
- Attach the canopy to the framework. If you are alone—

- Secure one edge of the folded canopy to the upper part of the pole nearest the intended entrance.
- Encircle the tripod with the canopy and attach the other edge to the upper part of the same pole.

If you have someone to help you, each of you should hold one edge of the folded canopy. Then—

- Place the center of the folded canopy on the tripod leg opposite the intended entrance.
- Walk in opposite directions around the tripod, meeting at the intended entrance.
- Secure the edges of the folded canopy to the upper part of the pole nearest the intended entrance.

If you intend to have a fire inside the tepee, you must leave a 12- to 20-inch opening at the top of the tepee for ventilation.

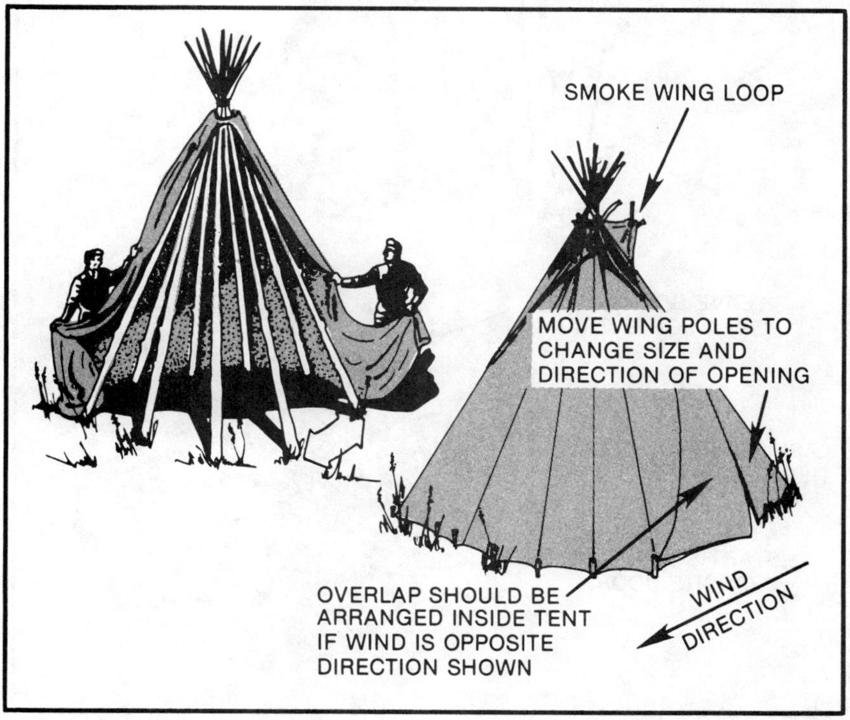

Figure 8-4. Parachute tepee.

Another type of parachute tepee requires a long pole and a sturdy tree. Use 14 gores of the chute. Attach the apex to a pole approximately 20 feet long. Scribe a 12-foot circle on the ground under a forked limb of the tree. Lay the pole in the fork of the limb. Stake down the parachute skirt on the circle.

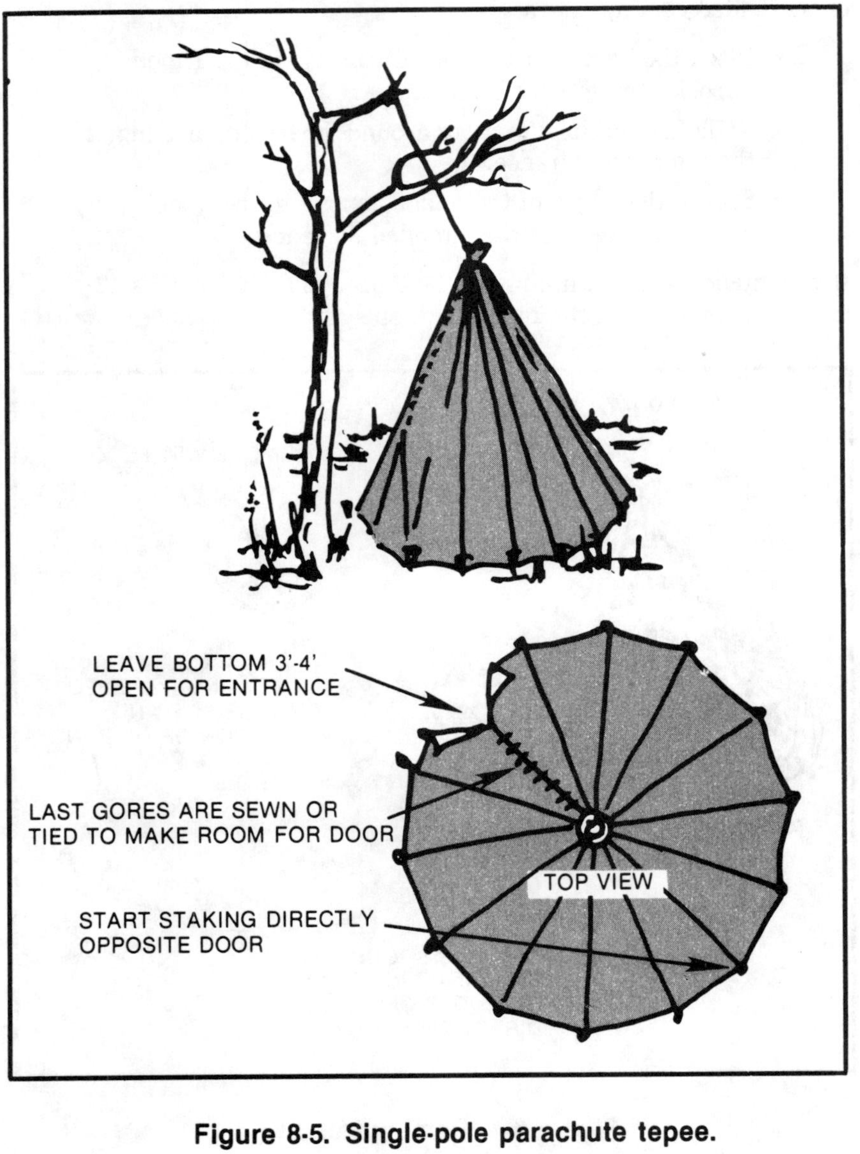

Figure 8-5. Single-pole parachute tepee.

A one-man shelter you can easily make using a parachute requires a tree and three poles. One pole should be about 15 feet long and the other two about 10 feet long.

To make this shelter (figure 8-6)—

- Secure the 15-foot pole to the tree at a point about waist high.
- Lay the two 10-foot poles on the ground on either side of and in the same direction as the 15-foot pole.
- Lay the folded canopy over the 15-foot pole so that about the same amount of material hangs on both sides.
- Tuck the excess material under the 10-foot poles, and spread on the ground inside the shelter to serve as a floor.
- Stake down or put a spreader between the two 10-foot poles at the shelter entrance so they will not slide inward.
- Use any excess material to cover the entrance.

The parachute cloth makes this shelter wind resistant, and the shelter is small enough that it is easily warmed. A candle, used carefully, can keep the inside temperature comfortable. This shelter is unsatisfactory, however, when snow is falling as even a light snowfall will cave it in.

Figure 8-6. One-man shelter.

Figure 8-7. Field expedient lean-to.

Field Expedient Lean-to. If you are in a wooded area and have sufficient natural materials, you can make an expedient lean-to (figure 8-7) without the aid of tools or with only a knife. You need more time to make it than the shelters previously mentioned, but it will protect you from most environmental elements.

You will need two trees (or two upright poles) about 6 feet apart; one pole about 7 feet long and 1 inch in diameter; five to eight poles about 10 feet long and 1 inch in diameter for beams; cord or vines for securing the horizontal support to the trees; and other poles, saplings, or vines to crisscross the beams.

To make this lean-to—

- Tie the 7-foot pole to the two trees at a point about waist to chest high. This is your horizontal support. (If there is a fork in the tree, you can rest the pole in it instead of tying the pole in place.) If a standing tree is not available, construct a bipod using Y-shaped sticks or two tripods.
- Place one end of the beams (10-foot poles) on one side of the horizontal support. As with all lean-to type shelters, make sure the backside of the lean-to is placed into the wind.
- Crisscross sapling or vines on the beams.
- Cover the framework with brush, leaves, pine needles, or grass, starting at the bottom and working your way up like shingling.
- Place straw, leaves, pine needles, or grass inside the shelter for bedding.

Figure 8-8. Swamp bed.

In cold weather you can add to the comfort of your lean-to by building a fire-reflector wall (figure 8-7). Drive four stakes about 4 feet long into the ground to support the wall. Stack green logs on top of one another between the support stakes. Bind the top of the support stakes so the green logs will stay in place. Fill in the spaces between the logs with twigs or small branches.

With just a little more effort you can have a drying rack: cut a few 3/4-inch diameter poles (length depends on distance between the lean-to support and the top of the fire-reflector wall). Lay one end of the poles on the lean-to horizontal support and the other ends on top of the reflector wall. Place and tie into place smaller sticks across these poles. You now have a place to dry clothes, meat, or fish.

Swamp Bed. At the beginning of the chapter we mentioned that a shelter is something that protects you from the elements. In a marsh or swamp, the swamp bed protects you from the wet ground. Factors that you must consider when selecting a site are weather, wind, tides, and available materials.

To make a swamp bed (figure 8-8)—

- Look for four trees clustered in a rectangle, or cut four poles (bamboo is ideal) and drive them firmly in the ground so they form a rectangle. They should be far enough apart and strong enough for your height and weight.
- Cut two poles long enough to span the width of the rectangle. They, too, must be strong enough to support your weight.

- Secure these two poles to the trees (or poles) high enough above the ground or water to allow for tides and high water.
- Cut additional poles long enough to span the length of the rectangle and lay them across the two short side poles.
- Cover the top of the bed frame with broad leaves or grass to form a soft sleeping surface.

Hammock. If you have a poncho, a parachute, or a shelter half, you can make a hammock (figure 8-9). If you have a nylon poncho, fold it in thirds before tying it off. This will give it the strength to support your weight. If you have a parachute, you can take four sections of the canopy, fold them to form a triangle, and then tie the triangle to trees.

Natural Shelters. Do not overlook natural formations that can give you shelter. Examples are caves, rocky crevices, clumps of bushes, small depressions, large rocks on leeward sides of hills, large trees with low hanging limbs, and fallen trees with thick branches. When selecting a natural formation, however—

- Stay away from low ground such as ravines, narrow valleys, or creek beds. Low areas collect the heavy cold air at night and are therefore colder than surrounding high ground. Thick, brushy, low ground also harbors more insects.

Figure 8-9. Hammock.

- Check for poisonous snakes, ticks, mites, scorpions, stinging ants, loose rocks that might fall on your shelter, and dead limbs, coconuts, or other natural growth that could fall on you during a storm.

Tree-pit Snow Shelter. If you are in a cold, deeply snow-covered area where evergreen trees grow and you have a digging tool, you can make a tree-pit snow shelter (figure 8-10).

To make this shelter—

- Find a tree with bushy branches that will give you overhead cover.
- Dig out the snow around the trunk of the tree until you reach the depth and diameter you desire or until you reach the ground.
- Pack the snow around the top and on the inside of the hole to provide support.
- Find and cut other evergreen boughs and place them over the top of the pit to give you additional overhead cover.

Figure 8-10. Tree-pit shelter.

Beach-shade Shelter. This shelter will protect you from the sun, wind, rain, and heat, and it is easy to make using natural materials.

To make this shelter (figure 8-11)—

- Find and collect driftwood or other natural material to use as support beams and to use as a digging tool.
- Select a building site that is above the high-water mark.
- Scrape or dig out a trench running north to south so that it receives the least amount of sunlight. Make the trench long enough and wide enough for you to lie down comfortably.
- Mound soil on three sides of the trench. The higher you make the mound, the more space you will have in your shelter.
- Lay support beams (driftwood or other natural material) on top of the mound spanning the trench to form the framework for a roof.
- Cover the framework with palm branches or other suitable material to form the roof.
- Enlarge the entrance to the shelter by digging out more sand in front of it.
- Use natural materials such as grass or leaves to form a bed inside the shelter.

Figure 8-11. Beach-shade shelter.

Desert Shelters. In an arid environment, you must consider time, effort, and material needed to make a shelter. If you have material such as canvas, parachute, poncho, or aircraft soundproofing, use it and a terrain feature as follows:

Rock method:

- Find an outcropped rock.
- Anchor one end of your poncho (canvas, parachute, or other material) on the edge of the outcropped rock using rocks or other weights.
- Extend and anchor the other end of the poncho so it gives you the best possible shade.

Mound method:

- Build a mound of sand or use the side of a sand dune for one side of the shelter.
- Anchor one end of the material (poncho, canvas, parachute) on top of the mound (or sand dune) using sand or other weights.
- Extend and anchor the other end of the material so that it gives you the best possible shade.

NOTE: *If you have sufficient material, fold it in half and form a 12-inch to 18-inch airspace between the halves. This will reduce the temperature below the shelter.*

Below-ground method. This type shelter (figure 8-12) reduces the midday heat as much as 30^0 to 40^0. Building it, however, requires more time and effort than building other type shelters. Since your physical effort will make you sweat more, increasing dehydration, you should wait until the cool of the night to make it.

To make this shelter—

- Find a low spot or depression between dunes or rocks, or if necessary, dig a trench 18 to 24 inches deep and long enough and wide enough for you to lie down comfortably.
- Pile the sand you take from the trench around three sides to form a mound.
- On the open end of the trench, dig out more sand so you can get in and out of your shelter easily.

- Cover the trench with material, such as parachute, poncho, or canvas.
- Secure the material in place using sand, rocks, or other weights.

If you have extra material, you can further decrease the midday temperature in the trench by securing the material 12 to 18 inches above the other cover. This layering of material will reduce the inside temperature 20° to 40°.

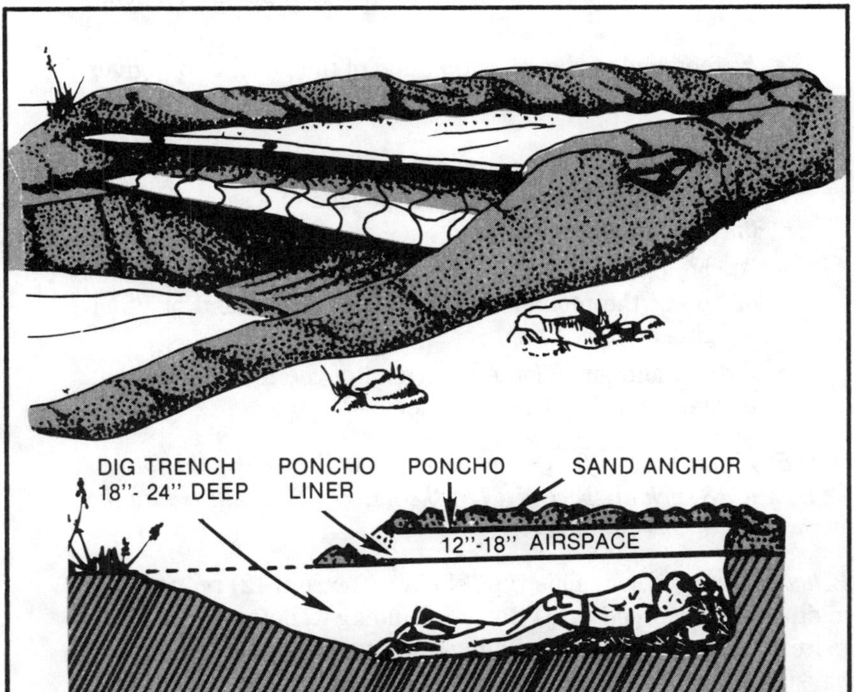

Figure 8-12. Desert shelter.

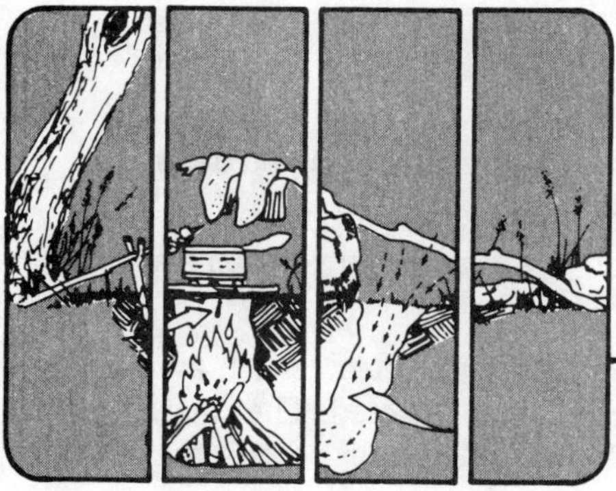

Chapter 9

Firebuilding

A fire can fulfill several needs: It can keep you warm; it can keep you dry; you can use it to cook food, to purify water, and to signal. But it can also cause problems when you are in enemy territory: It creates smoke, which can be smelled and seen from a long distance; it causes light, which can be seen day or night; it leaves signs of your presence.

Remember, you must always weigh your need for a fire against your need to avoid enemy detection.

When you are to operate in a remote area, you should always take a supply of matches in a waterproof case. Keep the matches on your person. Before going on such an operation, practice shielding a match flame in a fairly strong wind. Being able to keep a match aflame long enough to start a fire means you can save your other matches for future use.

Selecting and Preparing the Site

You will have to decide what site and arrangement to use for firebuilding based on—

- The area (terrain and climate) in which you are operating.
- The material and tools available.
- How much time you have.
- Why you need a fire.
- The nearness of the enemy.

Look for a dry spot—

- That is protected from the wind.
- That is suitably placed in relation to your shelter (if any).

- That will concentrate the heat in the direction you desire.
- Where a supply of wood or other fire-buring material is available. (See page 9-4 for types of material you can use.)

If you are in a wooded or brush-covered area, clear brush away and scrape the surface soil from the spot you selected. The cleared circle should be at least 3 feet (1 meter) in diameter so that there is little chance of the fire spreading.

If time allows, construct a fireplace wall using logs or rocks. This wall will help to reflect or direct the heat where you want it (figure 9-1). It will also reduce flying sparks and will cut down on the amount of wind blowing into the fire. However, you will need enough wind to keep the fire burning.

CAUTION: *Do not use wet or porous rocks as they may explode when heated.*

Figure 9-1. Types of firewalls.

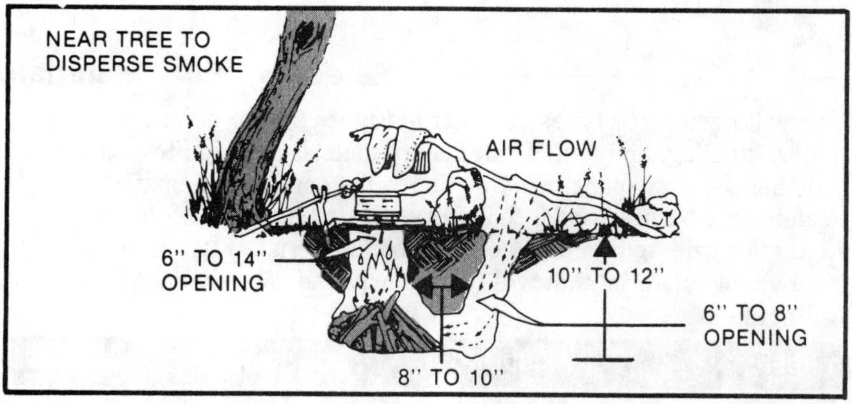

Figure 9-2. Dakota fire hole.

In some situations you may find that an underground fireplace will best meet your need. It conceals the fire to some extent and serves well for cooking food. To make an underground fireplace or Dakota fire hole (figure 9-2)—

- Dig a hole in the ground.
- On the upwind side of this hole, poke one large connecting hole for ventilation.
- Build your fire in the hole as illustrated.

If you are in a snow-covered area, you can use green logs to make a dry base for your fire (figure 9-3). Trees with wrist-size trunks are easily broken in extreme cold. Cut or break several green logs and lay them side by side on top of the snow. Add one or two more layers, laying the top layer logs in a direction opposite those on the layer below it.

Figure 9-3. Base for fire in snow-covered area.

Selecting Fire Material

You will need three types of material (figure 9-4) to build a fire: tinder, kindling, and fuel. Tinder is dry material that ignites with little heat—just a spark. Kindling is readily combustible material to add to the burning tinder to increase the temperature of the fire so that it will ignite less combustible material. Fuel is less-combustible material that will burn slowly and steadily once it is well ignited.

TINDER*	KINDLING*	FUEL
Birch bark	Small twigs	Dry standing wood and dry dead branches
Shredded inner bark from cedar, chestnut, red elm trees	Small strips of wood	Dry inside (heart) of fallen tree trunks and large branches
Fine wood shavings	Split wood	Green wood that is finely split
Dead grass, ferns, moss, fungi	Heavy cardboard	Dry grasses twisted into bunches
Straw	Pieces of wood removed from the inside of larger pieces	Peat dry enough to burn (this may be found at the top of undercut banks)
Sawdust	Wood that has been soaked or doused with highly flammable materials such as gasoline, oil, or wax	Dried animal dung
Very fine pitch-wood scrapings		Animal fats
Dead evergreen needles		Coal, oil shale, or oil sand lying on the surface
Punk (the completely rotted portion of dead logs or trees)		
Evergreen tree knots		
Bird down (fine feathers)		
Down seed heads (milkweed, dry cattails, bulrush, Canada thistle, goldenrod, dandelion)		
Fine, dried vegetable fibers		
Spongy threads of dead puffball		
Dead palm leaves		
Skinlike membrane lining bamboo		
Lint from pocket and seams		
Charred cloth		
Waxed paper		
Outer bamboo shavings		
Gunpowder		
Cotton		
Lint		*Must be completely dry.

Figure 9-4. Materials for building fires.

There are several methods for laying a fire for quick firemaking. Three easy methods are cone, lean-to, and cross-ditch.

Cone (figure 9-5). Arrange tinder and a few sticks of kindling in the shape of a cone. Fire the center. As the core burns away, the outside logs will fall inward, feeding the heart of the fire. This type of fire burns well even with wet wood.

Lean-to (figure 9-5). Push a green stick into the ground at a 30° angle. Point the end of the stick in the direction of the wind. Place some tinder (at least a handful) deep inside this lean-to stick. Lean pieces of kindling against the lean-to stick. Light the tinder. As the kindling catches fire from the tinder, add more kindling.

Cross-ditch (figure 9-5). Scratch a cross about 1 foot in size in the ground. Dig the cross 3 inches deep. Put a large wad of tinder in the middle of the cross. Build a kindling pyramid above the tinder. The shallow ditch allows air to sweep under the fire lay to provide a draft.

Your situation and the material available in the area may make another method more suitable:

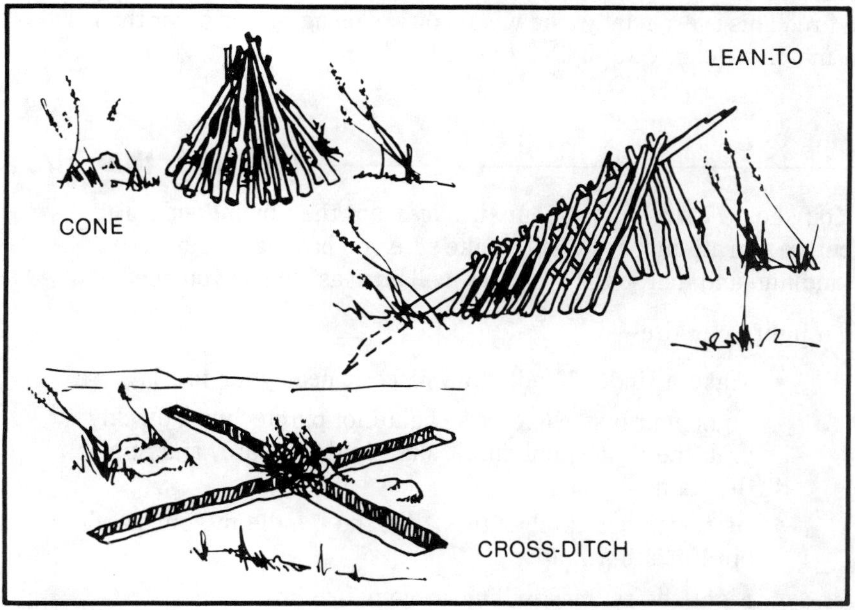

Figure 9-5. Methods for laying a fire.

Pyramid. Place two small logs or branches parallel on the ground. Place a solid layer of small logs or branches across the parallel logs. Add three or four more layers of logs or branches, laying each layer vertical to and making it smaller than the layer below it. Make a starter fire of tinder and kindling atop the pyramid. As the starter fire burns, it will ignite the logs below it. This will give you a fire that burns downward, radiates heat evenly all around, and requires no attention during the night.

Fire stick. Place two rocks on the ground about 10 inches apart. (If no rocks are available, stick two thick sticks or forked sticks in the ground.) Lay the fire stick cross the two rocks. Place a large handful of tinder under the fire stick. Lean kindling on the lee side of the fire stick.

Fire pit. Dig a 4- to 5-inch diameter hole 10 to 12 inches deep. Dig another hole 18 to 24 inches from the first hole to serve as the chimney. Dig a tunnel connecting the two holes. Use dry twigs only for fuel, keeping the fire as small as possible. You can use this method in a clandestine situation.

Keep in mind that good and sufficient tinder is essential in lighting a fire. This is especially true when you are using means other than a match or cigarette lighter to light the fire.

Lighting the Fire

You should always light your fire lay from the upwind side. But before you attempt to light it, make sure you have laid your tinder, kindling, and fuel so that your fire will burn as long as you need it.

To light your fire—

- Make a tinder bundle and place it near your fire lay.
- Light your match (cigarette lighter or pyrotechnic); quickly grab the tinder, holding it slightly downward, and apply the flame.
- As the tinder catches fire, turn the bundle in your hand until it is burning well.
- Light the tinder/kindling in your fire lay.

Don't waste your matches, your lighter fuel, or your pyrotechnics.

Light a candle if you have one, since it will burn longer. Save your matches and lighter fuel for later use.

If you lack matches, a cigarette lighter, or pyrotechnics, you can use other things (figure 9-6) to light a fire. Remember, though, you need good tinder and a good fire lay.

Convex Lens. This method is limited to bright sunlight days, and it takes time. The lens from your binoculars, camera, or telescopic sight, or the magnifying glass from your compass are suitable for this method. Angle the convex lens until the sun's rays are concentrated on the tinder. Hold the lens so that the sun rays stay directed over the same spot until the tinder begins to smolder. Gently fan or blow the tinder into a flame, and proceed as above.

Metal Match. Arrange some tinder in the shape of a bird's nest and place a dry leaf, thin dry bark, or a piece of paper in the tinder nest. Place the tinder near your fire lay. Holding your knife or similar item in one hand and the metal match in the other, place the metal match point on the dry leaf (bark, paper) in the center of the tinder nest. Scrape your knife against the metal match as if you are whittling. The sparks will hit the scrapings and tinder.

Battery. You may be able to strike a spark using a battery. The technique, however, depends upon the type of battery you have. With a vehicle battery attach a wire to each terminal. Set the battery where the wires will reach the tinder. Touch the bare ends of the wires together next to the tinder so the sparks will ignite it.

Flint and Steel. If you have a piece of steel—a knife blade, for instance—and a piece of flint, you can strike a spark to light tinder. (You can substitute a piece of hard stone for flint.) Hold the flint as near the tinder as possible. Strike the piece of steel downward on the flint so the sparks will hit the center of the tinder.

Bow and Drill. This is a last-resort method to use to light a fire. You will need—

- A piece of hardwood 2 to 3 feet long and 1 to 1½ inches in diameter for the bow.
- A piece of dry hardwood 6 to 9 inches long and ½ to ¾ inch in diameter for the drill.

- A piece of hardwood about 2" x 2" x 2" for the drill cap (a shell or stone will also work).
- A piece of dry softwood about 6" x 18" x 1" (the width and length can vary).
- A cord for the bowstring (a shoelace or thong will do).

To set up your bow and drill—

- String both ends of the bow loosely.
- Shape the drill so it is slightly tapered at the cap end and gently rounded at the drilling end.
- Carve out the center of the cap so the tapered end of the drill will fit inside it.
- Place the piece of softwood stably on the ground.
- Make a dent near the edge of the softwood to receive the drill.
- Make a notch that connects with the dent in the softwood.
- Place tinder next to the notch.
- Vigorously spin the drill, maintaining pressure on the cap so that there is friction between the softwood and drill. This friction produces the coals to start your fire.

Fire Saw. This, too, is a last-resort method. It consists of rubbing two pieces of wood vigorously against each other. Soft wood or split bamboo makes a good fire saw. When using bamboo—

- Select bamboo that is 2 to 3 inches in diameter.
- Cut a piece 2 to 3 feet long and split it lengthwise.
- Cut a notch on each edge of one split piece of bamboo so that the notches are aligned.
- Place tinder in the bamboo at the notches.
- Brace the bamboo, split side up, on the ground with a forked stick.
- With the other piece of split bamboo, saw in the notches until the tinder ignites.

Helpful Hints

Collect good tinder whenever you find it; carry it with you in a waterproof container.

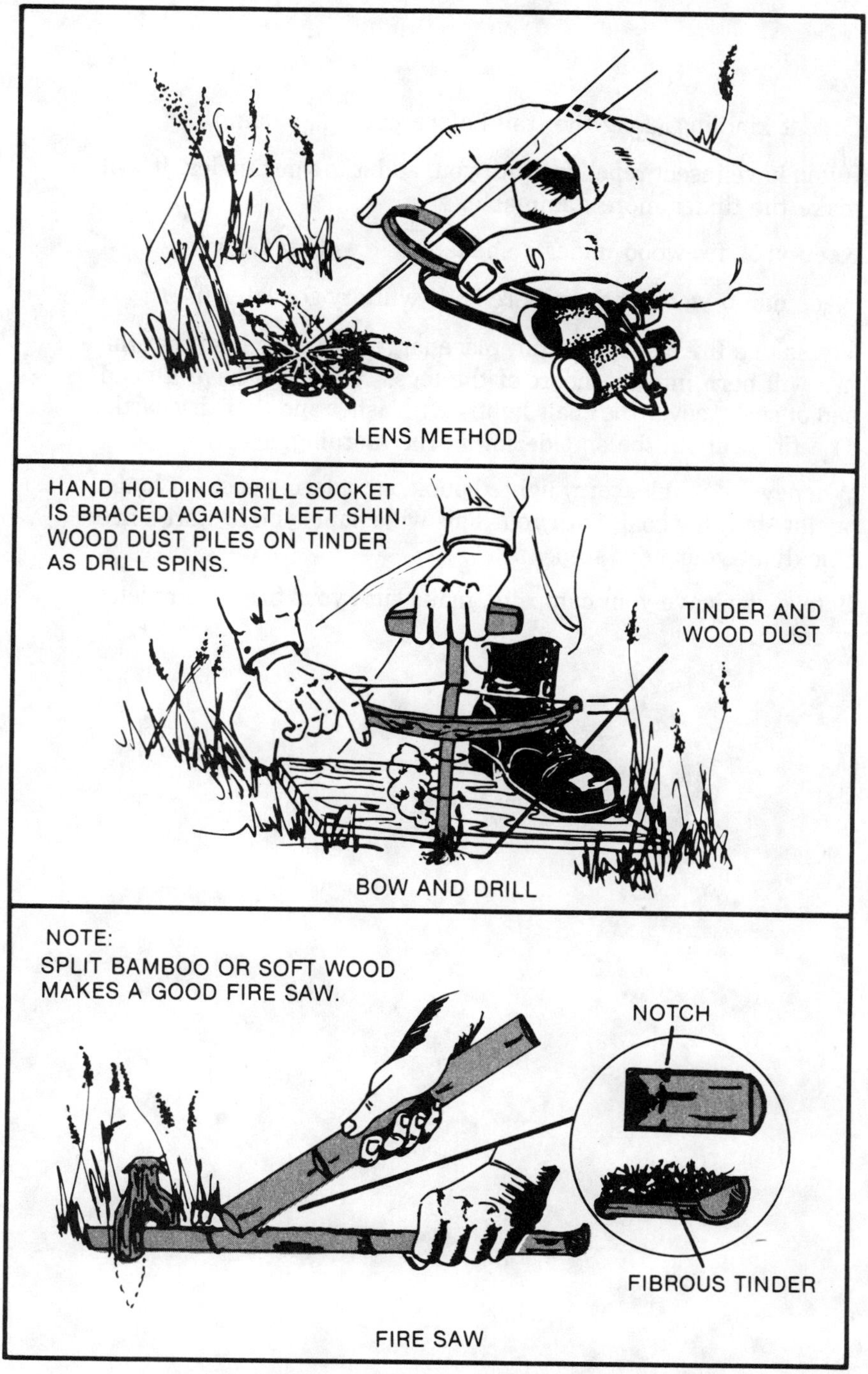

Figure 9-6. Fire making without matches.

Collect kindling along the trail before you make camp.

If you have insect repellent with you, add it to your tinder. It will make the tinder more flammable.

Keep your firewood under a shelter so it will stay dry.

Place damp wood near the fire so it will dry for later use.

To make a fire last overnight, place large logs over it so that the fire will burn into the heart of the logs. When you have a good bed of coals, cover the coals lightly with ashes and then dry earth. The fire will still be smoldering in the morning.

Whenever possible, carry lighted punk, smoldering coconut husk, or slow-burning coals with you. This will enable you to start a fire quickly at your next stop.

Before you leave your campsite, make sure your fire is completely out.

Chapter 10

Water Crossings

When you are in a survival situation in any area except the desert, you are likely to encounter a water obstacle. It may be in the form of a river, a stream, a lake, a bog, quicksand, quagmire, or muskeg. Whatever it is, you need to know how to cross it safely.

Rivers and Streams

A river or stream may be narrow or wide, shallow or deep, slow moving or fast moving. It may be snow-fed or ice-fed. Your first step is to find a place where the river is basically safe for crossing. So look for a high place from which you can get a good view of the river and find a place for crossing. If there is no high place, climb a tree. Check the river carefully for the following:

- A level stretch where it breaks into a number of channels. Two or three narrow channels are usually easier to cross than a wide river.
- Obstacles on the opposite side of the river that might hinder your travel. Try to select the spot from which travel will be the safest and easiest.
- A ledge of rocks that crosses the river. This often indicates dangerous rapids or canyons.
- A deep or rapid waterfall or a deep channel. Never attempt to ford a stream directly above or even close to such spots.
- Rocky places. Avoid these; you can be seriously injured from falling on rocks. An occasional rock that breaks the current, however, may assist you.

- A shallow bank or sandbar. If possible, select a point upstream from a shallow bank or sandbar so that the current will carry you to it if you lose your footing.
- A course across the river that leads downstream so that you will cross the current at about a 45^0 angle.

The depth of a fordable river or stream is no deterrent if you can keep your footing. In fact, deep water sometimes runs more slowly and is therefore safer than fast-moving shallow water. You can always dry your clothes later, or if you prefer, you can make a raft to carry your clothing and equipment across the river.

Rapids

Crossing a deep, swift river or rapids is not as dangerous as it looks. If you are swimming across, swim with the current—never fight it—and try to keep your body horizontal to the water. This will reduce the danger of being pulled under.

In fast, shallow rapids, go on your back, feet first; fin your hands alongside your hips to add buoyancy and to fend submerged rocks. Keep your feet up to avoid getting them bruised or caught by rocks.

In deep rapids, go on your belly, head first; angle toward shore whenever you can. Breathe between wave troughs. Be careful of backwater eddies and converging currents as they often contain dangerous swirls. Avoid bubbly water under falls; it has little buoyancy.

If you are going to ford a swift, treacherous stream, remove your pants and underpants so that the water will have less grip on your legs. Keep your shoes on to protect your feet and ankles from rocks and to give you firmer footing.

Tie your pants and important articles securely to the top of your pack. This way, if you have to release your pack, all your articles will be together. It is easier to find one large pack than to find several small items.

Carry your pack well up on your shoulders so you can release it quickly if you are swept off your feet. Not being able to get a pack off quickly enough can drag even the strongest of swimmers under.

Find a strong pole about 5 inches in diameter and 7 to 8 feet long to help you ford the stream. Grasp the pole and plant it firmly on your upstream side to break the current. Plant your feet firmly with each step, and move the pole forward a little downstream from its previous position, but still upstream from you. With your next step, place your foot below the pole. Keep the pole well slanted so that the force of the current keeps the pole against your shoulder (figure 10-1).

If there are other people with you, cross the stream together (figure 10-2). Make sure that everyone has prepared their pack and clothing as above. Have the heaviest person get on the downstream end of the pole and the lightest person on the upstream end. This way, the upstream person will break the current, and the persons below can move with comparative ease in the eddy formed by the upstream person. If the upstream person is temporarily swept off his feet, the other persons can hold steady while he regains footing.

As in all fording, cross the stream so that you will cross the downstream current at a 45° angle. Currents too strong for one person to stand against can usually be crossed safely in this manner.

Do not be concerned about the weight of your pack as the weight will help rather than hinder you in fording the stream. Just make sure you can release the pack quickly if necessary.

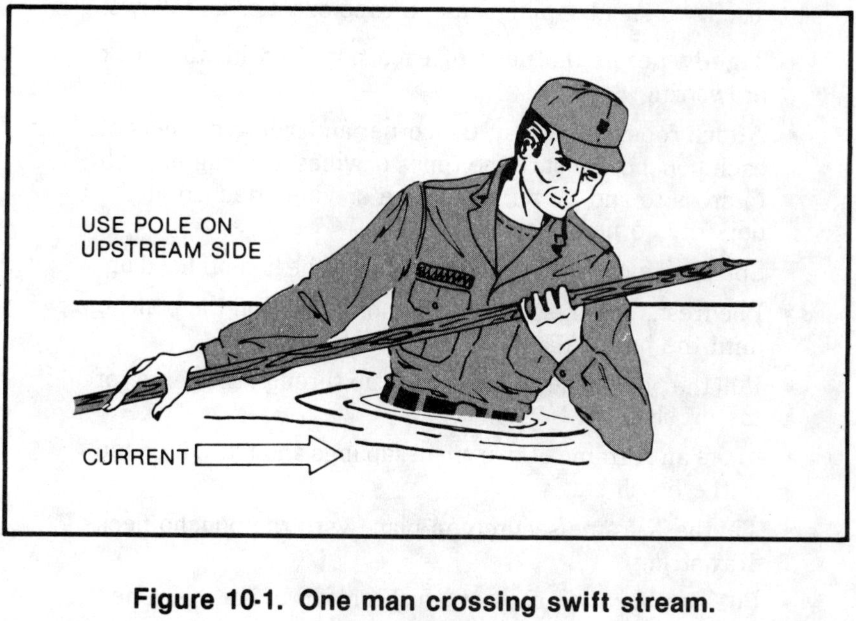

Figure 10-1. One man crossing swift stream.

Figure 10-2. Several men crossing swift stream.

Rafts

If you are with a buddy and each of you has a poncho, you can construct a brush or Australian poncho raft. With this type raft, you can safely float your equipment across a slow-moving stream or river.

Brush Raft. The brush raft (figure 10-3) will support about 250 pounds if properly constructed. To construct it, use ponchos; fresh, green brush; two small saplings; and a rope or vines as follows:

- Tightly tie off the neck of each poncho with the neck drawstring.
- Attach ropes or vines at the corner and side grommets of each poncho. Be sure the ropes or vines are long enough to cross to and tie with the rope or vine attached at the opposite corner or side.
- Spread one poncho on the ground with the tied-off hood up.
- Pile fresh, green brush (no thick branches) on the poncho until the brush stack is about 18 inches high.
- Pull the poncho neck drawstring up through the center of the brush stack.
- Make an X-frame of two small saplings and place it on top of the brush stack.
- Tie the X-frame securely in place with the poncho neck drawstring.
- Pile another 18 inches of brush on top of the X-frame.

- Compress the brush slightly.
- Pull the poncho sides up around the brush, and using the ropes or vines attached to the corner and side grommets, tie diagonally from corner to corner and from side to side.
- Spread the second poncho, tied-off hood up, next to the brush bundle.
- Roll the brush bundle onto the center of the second poncho so that the tied side is down.
- Tie the second poncho around the brush bundle in the same manner as you tied the first poncho around the brush.
- Tie one end of a rope to an empty canteen and the other end to the raft. This will aid you in towing.

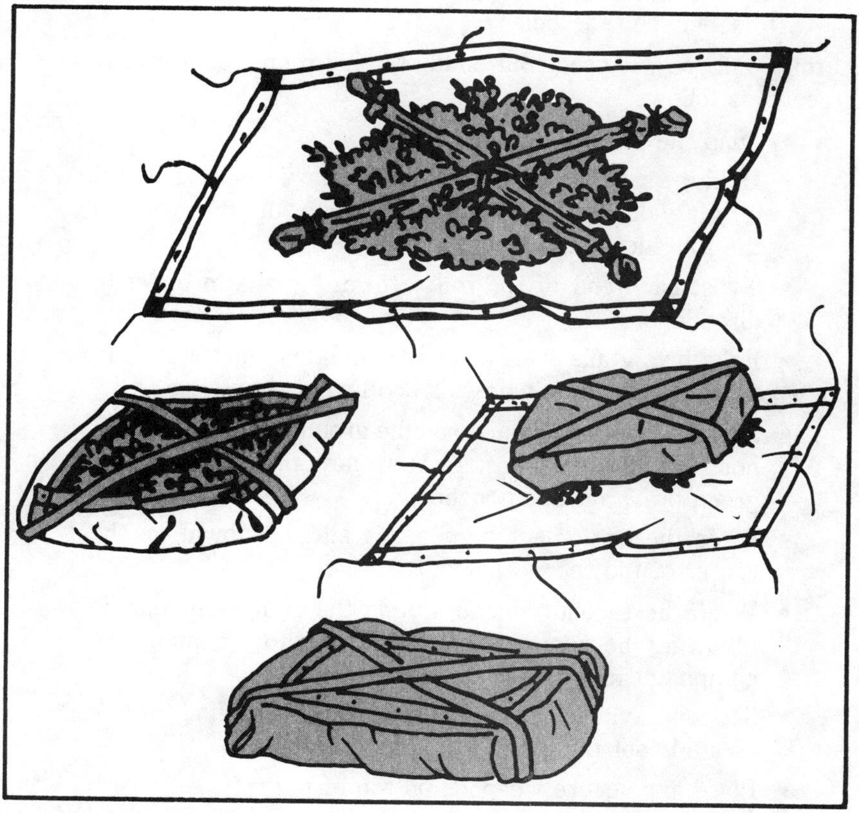

Figure 10-3. Brush raft.

Australian Poncho Raft. If you do not have time to gather brush for a brush raft, you can make an Australian poncho raft (figure 10-4). This raft, although more waterproof than the poncho brush raft, will only float about 80 pounds of equipment. To construct this raft, use two ponchos, two 4-foot poles or branches, and ropes, vines, bootlaces, or comparable material as follows:

- Tightly tie off the neck of each poncho with the neck drawstring.
- Spread one poncho on the ground with the tied-off hood up.
- Place and center the two 4-foot poles about 18 inches apart on the poncho.
- Place the rucksacks, packs, and/or other equipment between the poles. You should also place other items, such as your boots and outer garments, that you want to keep dry between the poles.

At this point you will need your buddy's help to complete the raft. Proceed as follows:

- Snap the poncho sides together.
- Hold the snapped portion of the poncho in the air and roll it tightly down to the equipment. Make sure that you roll the full width of the poncho.
- Twist each end of the roll to form pigtails in opposite directions.
- Fold the pigtails over the bundle and tie them securely in place using ropes, vines, or bootlaces.
- Spread the second poncho on the ground with the tied-off hood up. If you need more buoyancy, place some fresh, green brush on this poncho.
- Place the equipment bundle, pigtail side down, on the center of the second poncho.
- Wrap the second poncho around the equipment bundle following the same procedure you used for wrapping the equipment in the first poncho.
- Tie ropes, vines, or other binding material around the raft about 1 foot from the end of each pigtail.
- Place and secure weapons on top of the raft.
- Tie one end of a rope to a canteen and the other end to the raft. This will help you in towing the raft.

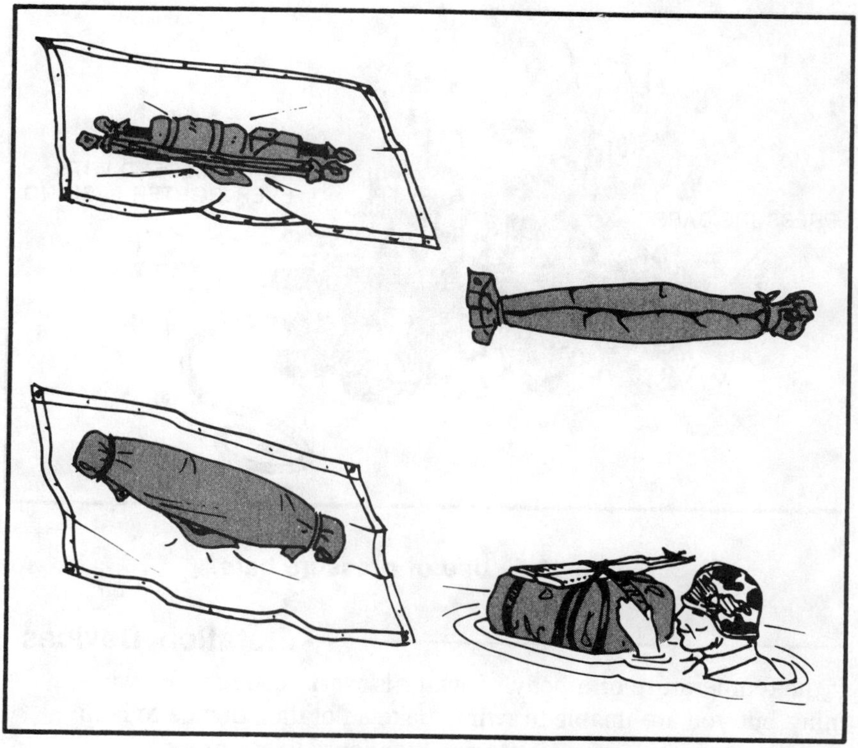

Figure 10-4. Australian poncho raft.

When launching or landing either the brush or Australian type poncho raft, take care not to puncture or tear it by dragging it on the ground. Let the raft lay on the water a few minutes to ensure that it floats before you start to cross the river or stream. If the river is too deep to ford, push the raft in front of you while swimming.

Be sure to check the water temperature before trying to cross a river or water obstacles. If the water is extremely cold and you are unable to find a shallow fording place in the river, do not attempt to ford it. Devise other means for crossing. For instance, you might improvise a bridge by felling a tree over the river. Or you might build a raft large enough to carry you and your equipment. For this, however, you will need an axe, a knife, and a rope or vines. And you will need time.

Log Raft. You can make a raft using dry, dead, standing trees for logs. However, spruce trees found in polar and subpolar regions make the best rafts. A simple method for making a raft is to use pressure bars lashed securely at each end to hold the logs together.

Figure 10-5. Use of pressure bars.

Flotation Devices

If the temperature of a body of water is warm enough for swimming but you are unable to swim, make a flotation device to help you. Some things you can use for flotation devices are—

- Trousers. Knot each trouser leg at the bottom and button the fly. With both hands grasp the waistband at the sides and swing the trousers in the air to trap air in each leg. Quickly press the sides of the waistband together and hold it underwater so that the air will not escape. You now have water wings to keep you afloat as you cross the body of water.

NOTE: *These "water wings" may have to be reinflated several times when crossing a large body of water.*

- Empty containers. Lash together empty tins, gas cans, or boxes and use as water wings. You should not use this type flotation, however, except in a slow-moving river or stream.
- Plastic bags. Air fill two or more plastic bags and securely tie them together at the mouth.
- Poncho. Roll green vegetation tightly inside your poncho so that you have a roll at least 8 inches in diameter. Tie the ends of the roll securely. You can wear it around your waist or across one shoulder and under the opposite arm.

- Logs. Use a stranded drift log if one is available. Or find a log near the water to use for flotation. Be sure to test the log before starting to cross, however, as some tree logs, palm for example, will sink even when the wood is dead.
- Cattails. Gather stalks of cattails and tie them in a bundle 10 inches or more in diameter. The many air cells in each stalk causes a stalk to float until it rots. Test the cattail bundle to be sure it will support your weight before attempting to cross a body of water.

Other Water Obstacles

Other water obstacles that you may face are bogs, quagmire, muskeg, or quicksand. *Do not* try to walk across. Trying to lift your feet while standing upright will make you sink deeper. Try to bypass these obstacles. If you are unable to bypass them, you may be able to bridge them using logs, branches, or foliage.

Another way to get across is to lie face downward with your arms spread and swim or pull your way across. Be sure to keep your body horizontal.

In swamps, the areas that have vegetation are usually firm enough to support your weight. However, vegetation will usually not be present in open mud or water areas. If you are an average swimmer, however, you should have no problem swimming, crawling, or pulling your way through miles of bog or swamp.

Quicksand is a mixture of sand and water that forms a shifting mass. It yields easily to pressure and tends to suck down and engulf objects resting on its surface. It varies in depth and is usually localized. Quicksand generally occurs on flat shores, in silt-choked rivers with shifting water courses, and near the mouths of large rivers. If you are uncertain whether a sandy area is quicksand, toss a small stone on it. The stone will sink in quicksand.

Although quicksand has more suction than mud or muck, you can cross it in the same way you do a bog; flatten out, face downward, arms spread, and move slowly across.

You can give yourself more buoyancy by forming air pockets in your clothing. Tie your pants at the ankles to form air pockets in the legs. Blow your breath inside the front opening of your collar to form air pockets over your shoulders.

Some water areas you must cross may have underwater and floating plants that will make swimming difficult. However, you can swim through relatively dense vegetation if you keep calm and do not thrash about. Stay as near the surface as possible and use the breast stroke with shallow leg and arm motion. Remove the plants around you as you would clothing. When you get tired, float or swim on your back until you are rested enough to continue with the breast stroke.

The mangrove swamp is another type of obstacle that occurs along tropical coastlines. Mangrove trees or shrubs throw out many prop roots that form dense masses.

To get through a mangrove swamp, wait for low tide. If you are on the inland side, look for a narrow grove of trees and work your way seaward through these, or look for the bed of a waterway or creek through the trees and follow it to the sea. If you are on the seaward side, work inland along streams or channels.

Be on the lookout for crocodiles. They are often found along and in the shallow water. If there are any near you, leave the water and scramble over the mangrove roots.

A raft is the best means of crossing a large swamp area.

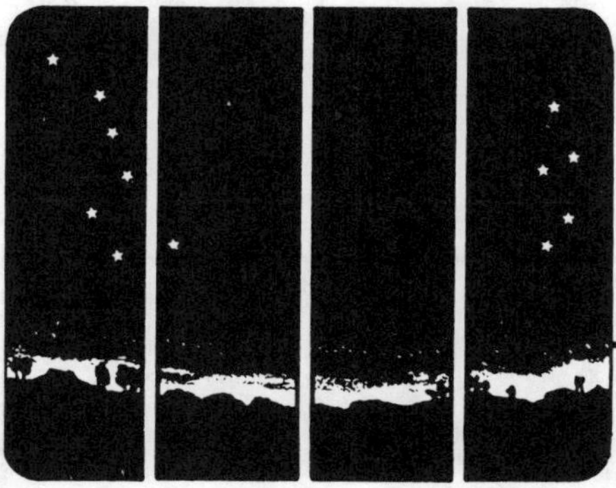

Chapter 11

Field Expedient Direction Finding

There are several methods by which you can determine direction by using the sun and the stars. These methods, however, will give you only a general direction. You can come up with a more nearly true direction if you have knowledge of the terrain of the territory or country. So it is up to you to learn all you can about the terrain, especially any prominent features or landmarks, of the country or territory to which you or your unit may be sent. This knowledge of the terrain together with the use of the methods explained below will enable you to come up with fairly true directions.

Using the Sun and Shadows

Two methods that are easy to use when there is sunlight are the shadow-tip and the watch.

Shadow-tip method. Everywhere on earth the sun rises in the east and sets in the west. Therefore, the following steps for the shadow-tip method (figure 11-1) are the same wherever you are:

- First, find a straight stick about 3 feet (1 meter) long and a fairly level, brush-free spot so that the stick will cast a definite shadow.
- Push the stick in the ground so it stands upright. It need not be perfectly vertical to the ground.
- Mark the tip of the shadow cast by the stick.
- Wait until the shadow moves 1½ to 2 inches (approximately 10 to 15 minutes).
- Mark the tip of the second shadow.
- Draw a line from the first mark through and about a foot beyond the second mark.

- Stand with your left foot on the first mark and your right foot on the end of the line you drew.

Figure 11-1. Shadow-tip method.

If you are in the northern temperate zone, you will be facing in a northerly direction and you will know the other directions by recalling their relation to north.

If you are in the southern temperate zone, you will be facing in a southerly direction.

Watch method. You can also determine direction using a watch (figure 11-2). The steps you take will depend on whether you are in the northern temperate zone or in the southern temperate zone and whether you have a conventional or digital watch. The northern temperate zone is located between 23.4⁰ north and 66.6⁰ north. The southern temperate zone is located between 23.4⁰ south and 66.6⁰ south.

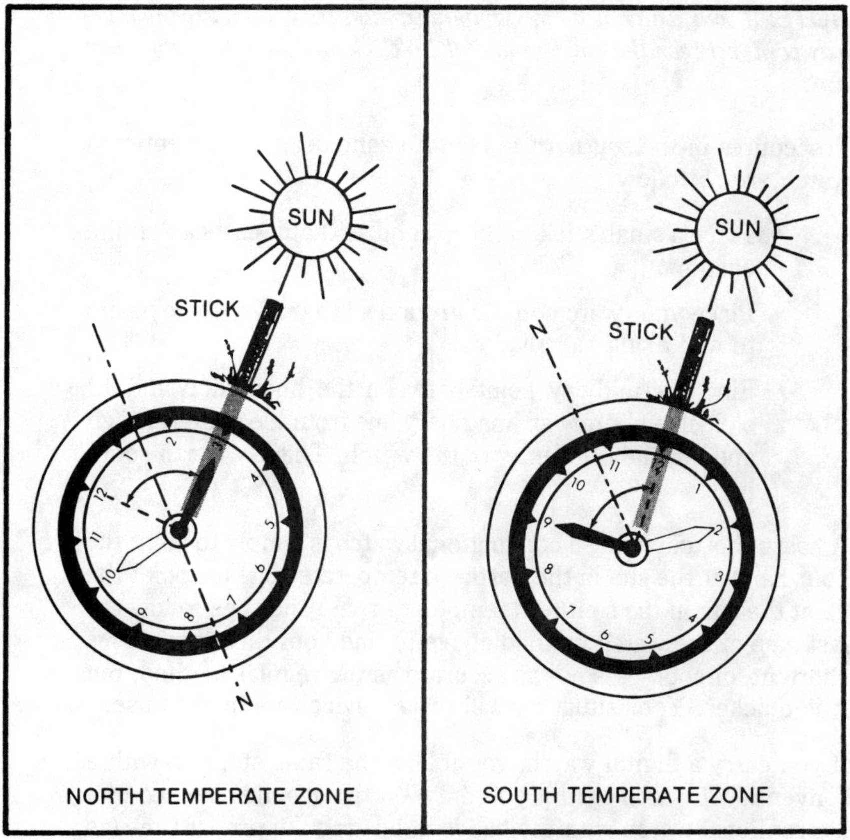

Figure 11-2. Watch method.

Procedures in the northern temperate zone using a conventional watch are as follows:

- Place a small stick in the ground so that it casts a definite shadow.
- Place your watch on the ground so that the hour hand points toward and along the shadow of the stick.
- Find the point on the watch midway between the hour hand and 12 o'clock and draw an imaginary line from that point through and beyond the center of the watch. This imaginary line is a north-south line. You can then tell the other directions.

NOTE: *If your watch is set on daylight savings time, then use the midway point between the hour hand and 1 o'clock to draw your imaginary line.*

Procedures in the southern temperate zone using a conventional watch are as follows:

- Place a small stick in the ground so that it casts a definite shadow.
- Place your watch on the ground so that 12 o'clock points to and along the shadow.
- Find the midway point between the hour hand and 12 o'clock and draw an imaginary line from the point through and beyond the center of the watch. This is a north-south line.

A hasty shortcut using a conventional watch is simply to point the hour hand at the sun in the northern temperate zone (or point the 12 at the sun in the southern temperate zone) and then follow the last step of the watch method above to find your directions. This shortcut, of course, is not as accurate as the regular method, but it is quicker. Your situation will dictate which method to use.

If you carry a digital watch, you follow the same steps as with a conventional watch except you draw on the ground the face of a conventional watch with the hands indicating the proper time (as shown on your digital watch) and the hour hand pointing to and along the shadow of the stick.

Using the Stars

On a clear night many stars are visible, and if you walk toward the North Star, you will be walking northward. The North Star, however, is not the brightest star in the sky and is sometimes hard to find. In order to locate the North Star, you should know that—

- All other stars revolve around the North Star.
- The North Star is the last star in the handle of the constellation Ursa Minor (Little Dipper), but the complete Little Dipper is often difficult to see.
- The easiest way to locate the North Star is by using the constellation Ursa Major (Big Dipper). A straight line drawn between the two stars (pointers) at the end of the Big Dipper's bowl will point to the North Star. The distance to the North Star is about five times the distance between the pointers.
- Directly across from the Big Dipper is the constellation Cassiopeia. It is made up of five stars and resembles a lopsided " M" or "W" depending on its position in the sky. The North Star is straight out from the center star of Cassiopeia. It is almost equidistant between the Big Dipper and the Cassiopeia.

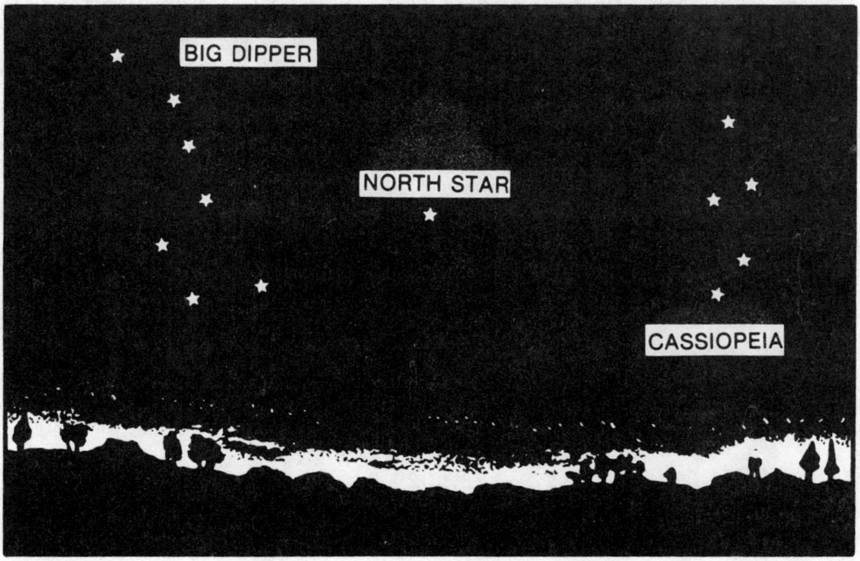

Figure 11-3. The Big Dipper.

South of the equator you can use the constellation Southern Cross to help you determine the general direction of south. The Southern Cross is a group of four bright stars in the shape of a cross that is tilted to one side. The two stars forming the long axis, or stem, of the cross are called pointers. To determine which direction is south—

- Imagine the long axis extends from its foot five times its length. The point where this imaginary line ends is in the general direction of south.
- Look straight down from this imaginary point to the horizon and select a landmark.

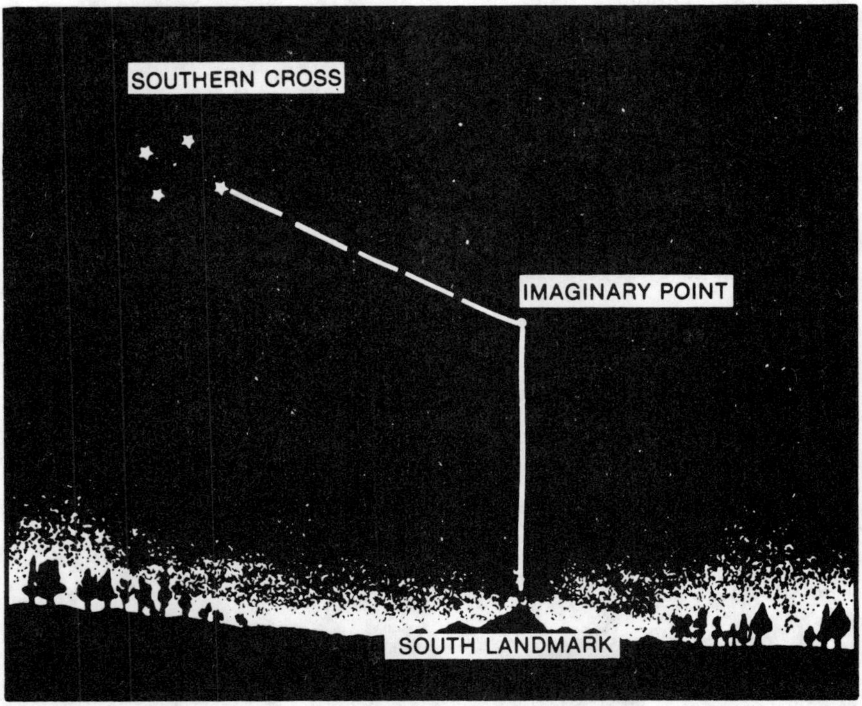

Figure 11-4. Southern Cross.

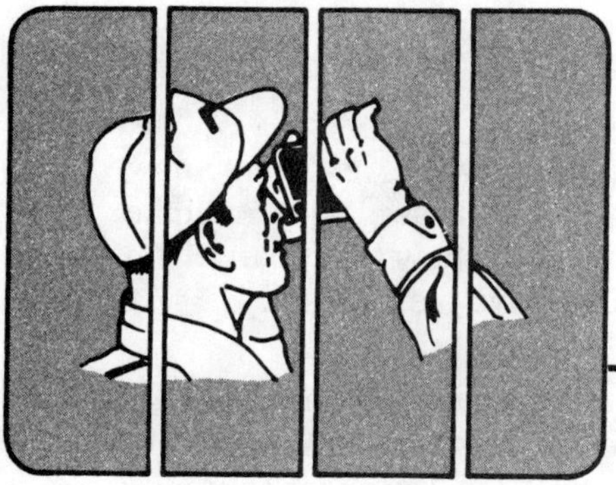

Chapter 12

Signaling

One of your first concerns when you find yourself in a survival situation is to communicate with your friends/allies. In a survival-evasion situation, you have the additional concern of avoiding the enemy.

Generally, communication is defined as "giving and receiving information." As a survivor, you must give signals in a manner that will make it easy for a rescue crew to receive the signal.

Prepare for Signaling

Your environment will have a bearing on the type of signals you can use. Keep in mind that from the air a pilot has difficulty spotting one man or a small group, especially when visibility is limited. So be prepared to use whatever signaling means will catch his attention no matter how high and how fast he is flying. However, be careful how you signal. In some circumstances the pilot could mistake the flashes of your signaling device as enemy ground fire.

Some attention-getters are man-made geometric patterns such as straight lines, circles, triangles, or X's in uninhabited areas; a large fire or flash of light; a large, bright object moving slowly; large signaling devices; or contrast, whether from color or shadows.

Whatever signaling technique or device you plan to use, know how to use it and be ready to use it on short notice. If possible, avoid the use of signals or signaling techniques that can endanger you. Keep in mind that signals to your friends/allies may alert the enemy of your presence and location. Before signaling, carefully weigh your chances of being rescued by friends against the danger of being captured by the enemy.

A radio is probably the surest and quickest way to let others know where you are and to let you receive their messages. So become familiar with the radios in your unit. Learn how to prepare them for operation and how to send and receive messages.

In the following paragraphs, you will find descriptions of other techniques, devices, and articles you can use for signaling. Learn how to use them. Think of ways in which you can adapt or modify them for different environments. And practice using these signaling techniques, devices, and articles before you need them.

Means for Signaling

There are several ways to signal. However, the means you use will depend on your situation and the material you have available.

Fire. During darkness, fire is the most effective means for signaling. Build three fires in a triangle (the international distress signal) or in a straight line with approximately 25 yards between the fires. Build them as soon as time and the situation permit and protect them until you need them. If you are alone, maintaining three fires may be difficult. If so, maintain one signal fire.

When constructing signal fires, consider your geographic location. For instance, if you are in the jungle, find a natural clearing or the edge of a stream where you can build the fires so that the jungle foliage will not hide them. You may even have to clear an area. If you are in a snow-covered area, you may have to clear the ground of snow or make a platform on which to build a fire so that melting snow will not extinguish it.

A burning tree is another means of attracting attention. Pitch-bearing trees can be set afire even when green. For other type trees, you can get the foliage to burn by placing dry wood in the lower branches and igniting it so that flames flare up and ignite the tree foliage. Cut and add more small green trees to the fire before the primary tree is consumed in order to produce more smoke. Always select a tree apart from other trees so that you do not start a forest fire and endanger yourself.

Smoke. During daylight, smoke is an attention-getter. The international distress signal is three columns of smoke. You should try to create a color of smoke that will contrast with the background: dark smoke against a light background and vice versa. If you practically smother a large fire with green leaves, moss, or a little water, the fire will produce white smoke. If you add rubber or oil-soaked rags to a fire, you will get black smoke.

In a desert environment smoke hangs close to the ground, but a pilot can spot it in open desert terrain.

Generally, smoke signals are effective only on comparatively calm, clear days. High winds, rain, or snow disperse smoke, lessening the chances of it being seen.

Smoke Grenades. If you have smoke grenades with you, you can use them in the same pattern as for fires. Be sure to keep your grenades dry so they will work when you need them. Take care not to ignite vegetation in the area when you use them.

M186 Pen Flare. This is a survival device that is included in an aviator's survival vest. It consists of a gun with a flare attached by means of a green nylon cord. When fired, the pen flare makes a sound like a pistol shot. The flare goes about 500 feet high and is about 10 feet in diameter.

To have the pen flare ready for immediate use, take it out of its wrapper, partially screw it in, leave it uncocked, and wear it on a cord or chain around your neck. Be ready to fire it in front of the aircraft and be ready with a secondary signal. Also be ready to take cover in case the pilot mistakes the flare for enemy fire.

Star Clusters. Red is the international distress color, so use a red star cluster whenever possible. Any color, however, will let your rescuers know where you are. Star clusters reach a height of 200 to 215 meters, burn on an average of 6 to 10 seconds, and descend at a rate of 14 meters per second.

Star Parachute Flares. These flares reach a height of 200 to 215 meters and descend at a rate of 2.1 meters per second. The M126 (red) burns for about 50 seconds and the M127 (white) for about 25 seconds. At night these flares can be seen for a distance of 30 to 35 miles.

Mirrors or Shiny Objects. On a sunny day, a mirror is your best signaling device. If you do not have a mirror, you can polish your canteen cup, your belt buckle, or a similar object so that it will reflect the sun's rays. Concentrate the flashes in one area so that they are secure from enemy observation. Practice using a mirror or shiny object for signaling *now;* do not wait until you need it.

If you have an MK-3 signal mirror, follow the instructions on its back.

Wear the signal mirror on a cord or chain around your neck so that you have it ready for immediate use. However, make sure the glass side is against your body so that it will not flash and be seen by the enemy.

CAUTION: *Do not flash a signal mirror rapidly as a pilot may mistake the flashes for enemy fire. Do not direct the beam on the cockpit of an aircraft for more than a few seconds as it may blind the pilot.*

Haze, ground fog, and mirages may make it hard for a pilot to spot signals from a flashing object. So, if possible, get to the highest point in your area when signaling. If you are unable to determine the aircraft's location, flash your signal in the direction of the aircraft noise.

NOTE: *Mirror signals can be seen over 70 miles under normal conditions in most environments and over 100 miles in a desert environment.*

Flashlight or Strobe Light. At night you can use a flashlight or a strobe light to send an SOS to an aircraft. When you use a strobe light, take care to prevent the pilot from mistaking it for incoming rounds. The strobe light flashes 60 times per minute. Some strobe lights have infrared covers and lenses.

VS-17 Panel. During daylight you can use a VS-17 panel to signal. Place the orange side upward as it is easier to see from the air than the violet side. Flashing the panel will make it easier for the pilot to spot.

Clothing. Spreading clothing on the ground or in the top of a tree is another way to signal. Select articles whose color will contrast

HOW TO USE THE MK-3 SIGNAL MIRROR

1. Reflect sunlight from mirror onto a nearby surface (raft, hand, etc.).
2. Slowly bring mirror up to eye-level and look through sighting hole. You will see a bright spot of light. This is the aim indicator.
3. Hold mirror near the eye and slowly turn and manipulate it so that the bright spot of light is on the target.
4. In friendly areas where only rescue by friendly forces is anticipated, free use of the mirror is recommended. Even though no aircraft or ships are in sight, continue to sweep the horizon. Mirror flashes may be seen for many miles, even in hazy weather. In hostile areas, the signal mirror must be used as an aimed signal only.

Figure 12-1. Mirror signals.

with the natural surroundings. Arrange them in a large geometric pattern so that they are more likely to attract attention.

Natural Material. If you lack other means, you can use natural materials to form a symbol or message that can be seen from the air. Build mounds that will cast shadows; you can use brush, foliage of any type, rocks, or snow blocks.

In snow-covered areas, tramp down the snow to form letters or symbols and fill in with contrasting material such as twigs or branches.

In sand, use boulders, vegetation, or seaweed to form a symbol or message.

In brush-covered areas, cut out patterns in the vegetation or sear the ground.

In tundra, dig trenches or turn the sod upside down.

In any terrain, use contrasting materials so that the symbols are visible to aircraft crews.

Sea Dye Markers. An Army aircraft that is intended for operations near or over water will normally carry a water survival kit that contains sea dye markers. If you are in a water survival situation, use sea markers during daylight to indicate your location. These spots of dye stay conspicuous for about 3 hours except in very rough seas, so you should use them only if you are in a friendly area. Keep the markers wrapped until you are ready to use them. Use them only when you hear or sight an aircraft. Sea dye markers are also very effective on snow-covered ground using distress code letters.

Codes/Signals

Now that you know how to let people know where you are, you need to know how to give them more information. It is easier to form one symbol than to spell out an entire message, so learn the following codes and symbols, which all aircraft pilots understand.

Morse Code. You can use lights or flags to send an SOS—three dots, three dashes, three dots—which is the internationally recognized signal of distress in radio code. Keep repeating the signal. When using flags, hold flags on the left side for dashes and on the right side for dots.

No.	Message	Code Symbol
1	Require Assistance	V
2	Require Medical Assistance	X
3	No or Negative	N
4	Yes or Affirmative	Y
5	Proceeding In This Direction	↑

Figure 12-2. Pattern signals.

Ground-to-Air Emergency Code. This code (figure 12-2) is actually five definite, meaningful symbols.

Body Signals. When an aircraft is close enough for the pilot to see you clearly, use body movements or positions (figure 12-3) to convey a message.

Figure 12-3. Close-in visual signals.

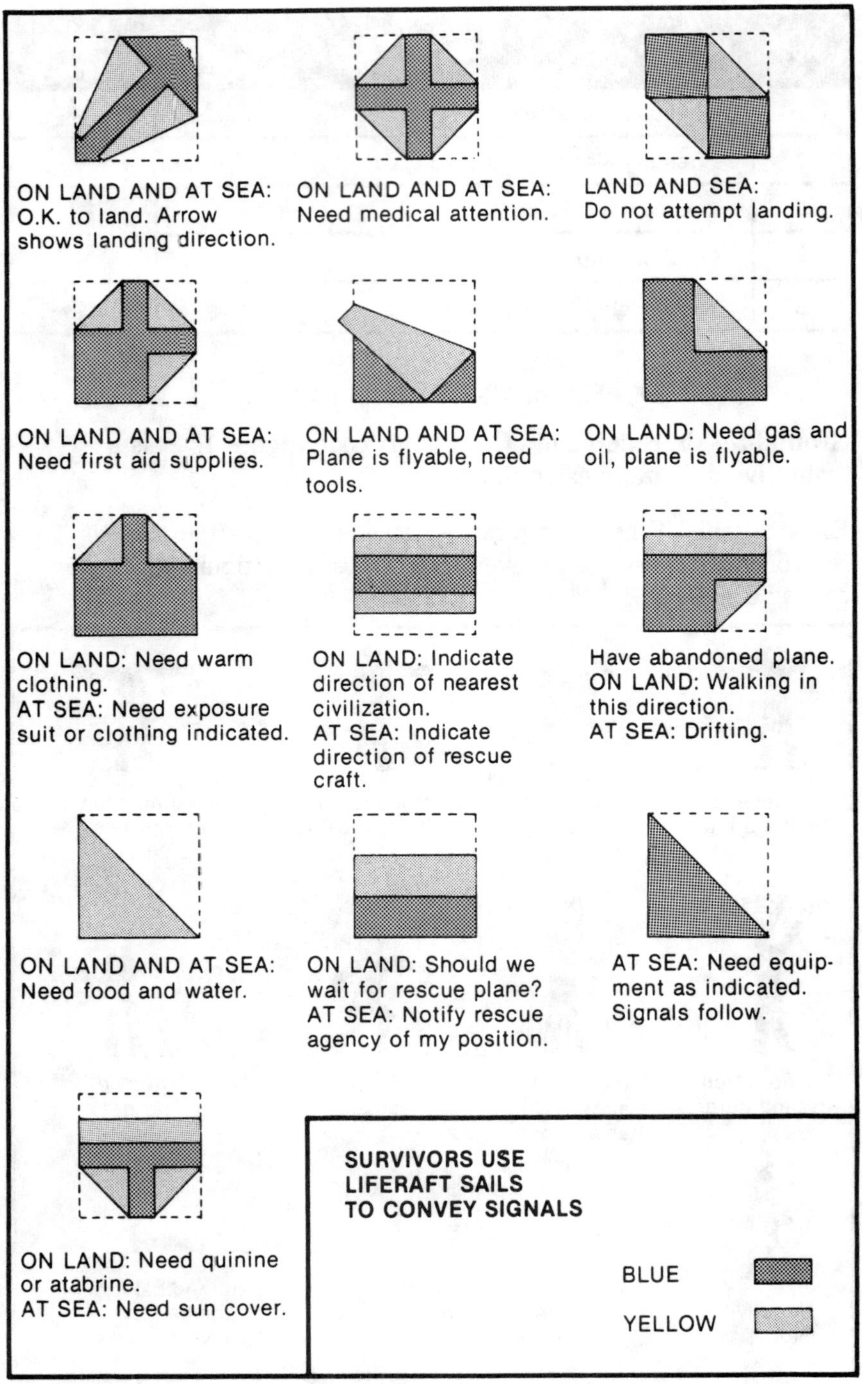

Figure 12-4. Panel signals.

Panel. If you have a VS-17 panel or a suitable substitute, use the symbols shown here (figure 12-4) to convey a message.

Aircraft Acknowledgements

The pilot of a fixed-wing aircraft will indicate he has seen you by moving the plane and flashing lights as shown in figure 12-5.

MESSAGE RECEIVED AND UNDERSTOOD

Aircraft will indicate that ground signals have been seen and understood by—

DAY OR MOONLIGHT: Rocking from side to side

NIGHT: Making green flashes with signal lamp

MESSAGE RECEIVED AND NOT UNDERSTOOD

Aircraft will indicate that ground signals have been seen but NOT understood by—

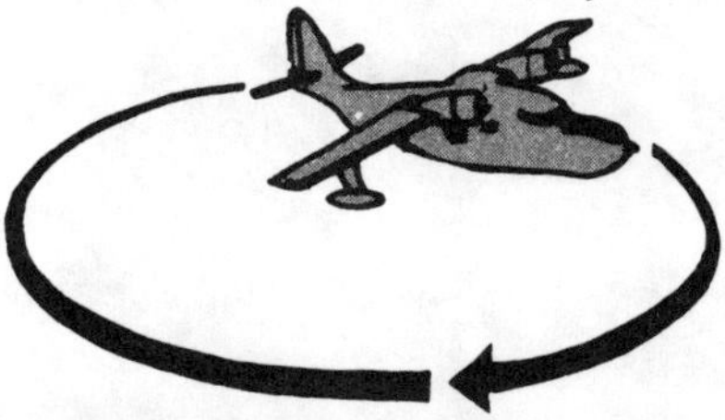

DAY OR MOONLIGHT: Making a complete circle

NIGHT: Making red flashes with signal lamp

Figure 12-5. Standard aircraft acknowledgements.

Be prepared to relay other messages to the pilot once he has acknowledged that he has received and understood your first message.

Chapter 13

Desert Survival

Survival in an arid area (a desert), as in any area, depends upon your knowledge of the terrain and the basic climatic elements, your ability to cope with them, and your will to live.

Terrain

Most arid areas have several types of terrain, each type seemingly blending into another type. The five basic types are:

- Mountainous.
- Rocky plateau.
- Sand dune.
- Salt marsh.
- Highly dissected terrain (called "gebel").

These types of terrain will not only affect your ability to procure water, food, and shelter (Chapters 5-8), they will also—

- Make physical movement quite demanding.
- Make land navigation difficult.
- Limit cover and concealment.

Mountain deserts are characterized by scattered ranges or areas of barren hills or mountains, separated by dry, flat basins. High ground may rise gradually or abruptly from flat areas to a height of several thousand feet above sea level.

Rainfall, although infrequent, occurs on high ground, runs off rapidly in the form of flash floods, erodes deep gullies and ravines, and deposits sand and gravel around the edges of the basins. Vegetation may appear after rain, but the water evaporates rapidly, leaving the land barren as before. If more water than can evaporate enters a basin, a shallow lake may develop. The Great Salt Lake in Utah and the Dead Sea are examples of such lakes. The water in such lakes has a high salt content and is undrinkable.

Mountain deserts at high altitudes have thin air and little or no vegetation. Sunburn is a danger. Climbing at high altitudes requires extra physical exertion and increases your need for water.

Movement on mountains during darkness is extremely dangerous.

Rocky plateau deserts are characterized by many solid or broken rocks at or near the surface. There may be sand dunes around the plateau. The extensive flat areas have relatively little relief, but the rock outcroppings offer shade as well as cover and concealment.

The rocks often form natural cisterns that collect water after rains. Look closely for these areas. Sometimes animal or bird indicators, such as trails, droppings, or birds in flight, may point out water sources.

Movement at night is dangerous.

Sand dune deserts are extensive areas covered with sand and gravel. The word "flat" is somewhat misleading in that some areas may have sand dunes over 1,000 feet (300 meters) high and 10 to 15 miles (16 to 24 kilometers) long. Other areas may be totally flat for distances of 2 miles (3.22 kilometers) or more. The area may be void of plant life or it may have scrub over 6 feet (1.82 meters) high.

The Sahara Desert, the Sinai Desert, the empty quarter of the Arabian Desert, the California and New Mexico deserts, and the Kalahari Desert in South Africa are examples of sand dune deserts.

You should avoid travel through sand dune deserts if possible.

Salt marshes are flat, desolate areas, sometimes studded with clumps of grass but devoid of other vegetation. They occur in arid areas where rainwater has collected, evaporated, and left large deposits of alkali salts and/or water with a high concentration of salts. The water is so salty it is undrinkable. A crust, which may be an inch to a foot thick, forms over the saltwater.

In arid areas there are salt marshes hundreds of kilometers square. These areas support large numbers of insects, most of which bite.

Avoid salt marshes. These types of terrain are highly corrosive to boots, clothing, and skin.

Highly dissected terrain is found in all arid areas. This terrain (called gebel or wadi) is formed by rainstorms that erode soft sand and carve out miniature canyons. A wadi may range from 10 feet (3 meters) wide and 6 feet (2 meters) deep to several hundred meters wide and deep. The direction it takes varies as much as its width and depth, twisting and turning, forming a maze. You can easily become lost because of this mazelike pattern. A wadi will give you good cover and concealment, but you should not try to move through it.

Environmental Factors

To survive and to evade an enemy in an arid area, you must know about and be prepared for the environmental conditions you face. You must determine the equipment you will need, the tactics you will use, and how the environment will impact on them and you.

In an arid area there are eight environmental factors that you must consider:

- Low rainfall.
- Intense sunlight and heat.
- Wide temperature range.
- Sparse vegetation.
- High mineral content near ground surface.
- Sandstorms.
- Mirages.
- Light levels.

Low rainfall is the most obvious environmental factor in an arid area. Some desert areas receive less than 4 inches of rain annually, and this comes in brief torrents that quickly run off the ground surface. With the high desert air temperatures you cannot survive long without water. So in a desert survival situation, you must first consider "How much water do I have?" and "Where are other sources?" (See Chapter 5.)

Intense sunlight and heat are present in all arid areas. Air temperature can rise as high as 140° during the day. Heat gain results from direct sunlight, hot blowing winds, reflective heat (the sun's rays bouncing off the sand), and conductive heat from direct contact with the desert sand and rock.

The temperature of desert sand and rock averages 30° to 40° more than that of the air. For instance, when the air temperature is 110°, the sand temperature may be 140°.

Intense sunlight and heat increase the body's need for water. To conserve your body sweat and energy, you need a shelter to reduce your exposure to the heat of the day. Travel at night to minimize the use of water. You can survey the area at dawn, dusk, or by moonlight when there is little likelihood of mirage.

Radios and sensitive items of equipment exposed to direct intense sunlight will malfunction.

Temperatures may get as high as 130° during the day and as low as 50° during the night in arid areas. The drop in temperature at night occurs rapidly and will chill a person who lacks warm clothing

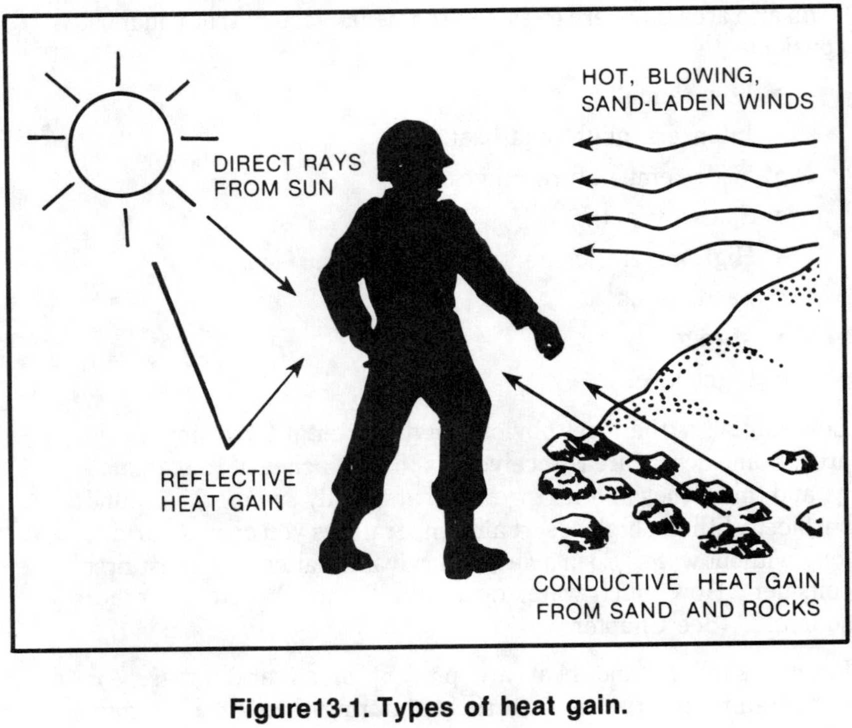

Figure13-1. Types of heat gain.

and is unable to move about. The cool evenings and nights are the best times to work or travel. If you plan to rest at night, you will find a wool sweater, long underwear, and a wool stocking cap extremely helpful.

Vegetation is sparse in arid areas. You will therefore have difficulty finding shelter and camouflaging your movements. During daylight hours large areas of terrain are visible and easily controlled by a small opposing force.

As an evadee, you should follow the principles of desert camouflage:

- Hide or seek shelter in dry washes (wadis) with thicker growths of vegetation and cover from oblique observation.
- Use the shadows cast from brush, rocks, or outcroppings. The temperature in shaded areas will be 20^{0} to 30^{0} cooler than the air temperature.
- Cover objects that will reflect the light from the sun.

Prior to moving, survey the area for sites that provide cover and concealment. A problem you will have is estimating distance. The emptiness of a desert terrain causes most people to underestimate distance by three: what appears to be 1 mile (1.61 kilometers) away is really 3 miles (4.83 kilometers) away.

All arid regions have areas where the surface soil has a high mineral content (borax, salt, alkali, and lime). Material in contact with this soil wears out quickly, and water in these areas is extremely hard and undrinkable. Wetting your uniform in such water to cool off may cause a skin rash. The Great Salt Lake area in Utah is an example of this type of mineral-ladened water and soil. There is little or no plant life; therefore, shelter is hard to find. Avoid these areas if possible.

Sandstorms (sand-laden winds) occur frequently in most deserts. The "Seistan" desert wind in Iran and Afghanistan blows constantly for up to 120 days. Within Saudi Arabia, winds average 2 to 3 miles per hour (mph) and can reach 70 to 80 mph in early afternoon. Major sandstorms and dust storms can be expected at least once a week.

The greatest danger is getting lost in a swirling wall of sand. You should wear goggles and cover your mouth and nose with cloth. If natural shelter is unavailable, mark your direction of travel, lie down, and ride out the storm.

Dust and wind-blown sand interfere with radio transmissions. Therefore, you should use other means for signaling, such as pyrotechnics, if available.

Mirages are optical phenomena caused by the refraction of light through heated air rising from a sandy or stony surface. They occur in the interior of the desert about 6 miles (9.6 kilometers) from the coast. They make objects that are 1 mile (1.6 kilometers) or more away appear to move.

This mirage effect makes it difficult for you to identify an object from a distance. It also blurs distant range contours so much that you feel as if you are surrounded by a sheet of water from which elevations stand out as "islands."

The mirage effect makes it hard for a person to identify targets, estimate range, and spot personnel. However, if you can get to high ground (10 feet or more above the desert floor), you can get above the superheated air close to the ground, overcoming the mirage effect.

Mirages make land navigation difficult because they obscure natural features.

Light levels in desert areas are more intense than in other geographic areas.

Moonlit nights are usually crystal clear; winds die down, haze and glare disappear, and visibility is good. You can see lights, red flashlights, and blackout lights great distances away. Noise carries far.

Conversely, during nights with little moonlight, visibility is extremely poor. Traveling is extremely hazardous; you must take care to avoid getting lost, falling into ravines, or stumbling into enemy positions. Movement during such a night is practical only if you have a compass and have spent the day in shelter, resting, observing and memorizing the terrain, and selecting your route.

Need For Water

The subject of man and water in the desert has incited considerable interest and confusion since the early days of World War II when the US Army was preparing to fight in North Africa. At one time the US Army and the Israeli Defense Forces thought they could condition men to do with less water by progressively reducing their water supplies during training. They called it water discipline. It caused hundreds of heat casualties.

A key factor in arid area survival is understanding the relationship between physical activity, air temperature, and water consumption. The body requires a certain amount of water for a certain level of activity at a certain temperature. For example, a man performing hard work in the sun at 110° F requires 5 gallons of water a day. Lack of the required amount of water causes a rapid decline in a person's ability to make decisions and to perform tasks efficiently (Chapter 2).

Your body's normal temperature is 98.6° F. Your body gets rid of excess heat (cools off) by sweating. The warmer your body becomes, whether caused by work, exercise, or air temperature, the more you sweat. The more you sweat, the more moisture you lose. Sweating is the principal cause of water loss. If a man stops sweating during periods of high air temperature and heavy work or exercise, he will have a heat stroke. This is an emergency that requires immediate medical attention.

Understanding how the air temperature and your physical activity affect your water requirements allows you to take measures to get the most from your water supply. These measures are:

- Find shade! Get out of the sun! Place something between you and the hot ground. Limit your movements!
- Conserve your sweat. Wear your complete fatigue uniform to include T-shirt; roll the sleeves down, cover your head, and protect your neck with a scarf or similar item. This will protect your body from hot blowing, sand-laden winds and the direct rays of the sun. Your clothing will absorb your sweat, keeping it against your skin so that you gain its full cooling effect. By staying in the shade quietly, fully clothed, not talking, keeping your mouth closed and breathing through your nose, your water requirement for survival drops dramatically.

- If water is scarce, do not eat. Food requires water for digestion; therefore, eating food will use water that you need for cooling.

Thirst is not a reliable guide for your need for water. A person who uses thirst as a guide will drink only two-thirds of his daily requirement. To prevent this "voluntary" dehydration, use this guide:

- At temperatures below 100^{0} F, drink 1 pint of water every hour.
- At temperatures above 100^{0} F, drink 1 quart of water every hour.

Drinking water at regular intervals helps your body to remain cool, decreasing sweating. Even when your water supply is low, sipping water constantly will keep your body cooler and reduce water loss through sweating. Conserve your sweat by reducing activity during the heat of day. DO NOT ration your water. If you attempt to ration water, you stand a good chance of becoming a heat casualty.

Heat Casualties

No matter how much instruction people receive on how to avoid becoming overheated, some heat casualties will occur. Following are the major types of heat casualties and their treatment *when little water and no medical assistance are available:*

- Heat cramps. These are caused mainly by loss of salt due to excessive sweating. Symptoms are moderate to severe muscle cramps in legs, arms, or abdomen. These symptoms may start as mild muscular discomfort. This is the time when the person should stop all activity, get in the shade, and drink water. If the person fails to recognize the early symptoms and continues with physical activity, he will have severe muscle cramps and pain. Treat the person the same as for heat exhaustion, below.
- Heat exhaustion. This is caused by a large loss of body water and salt. Symptoms are headache; mental confusion; irritability; excessive sweating; weakness; dizziness;

cramps; and pale, moist, cold (clammy) skin. Take the following steps immediately: Get the patient under shade, make him lie on a stretcher or similar item approximately 18" off the ground, loosen his clothing, sprinkle him with water, fan him, and have him drink small amounts of water every 3 minutes. Ensure that he remains quiet and rests.

- Heat stroke. This is a severe heat injury caused by extreme loss of water and salt and the body's inability to cool itself. The patient may die if not cooled immediately. Symptoms are no sweating; hot, dry skin; headache; dizziness; fast pulse; nausea and vomiting; and mental confusion leading to unconsciousness. Take the following steps immediately: Get the person to shade, lay him on a stretcher or similar item approximately 18" off the ground; loosen his clothing; pour water on him (it does not matter if the water is polluted or brackish); fan him; massage his arms, legs, and body. If he regains consciousness, let him drink small amounts of water every 3 minutes.

Precautions

In a desert survival/evasion situation, it is unlikely that you will have a medic or medical supplies with you to treat heat injuries, so you should take extra care to avoid heat injuries. Rest during the day; do your work during the cool evenings and night. Use a buddy system to watch for heat injury, and follow these guidelines:

- Make sure you tell your buddy where you are going and when you will return.
- Watch your buddy for signs of heat injury. If he complains of being tired or wanders away from the group, he may be a heat casualty.
- Make sure your buddy drinks water at least once an hour.
- Get in the shade when resting; do not let your buddy lie directly on the ground.
- Do not let your buddy take off his shirt and work during the day.
- Check the color of your urine; a light color means you are

drinking enough water; a dark color means you need to drink more.

By looking at figure 13-2, you can determine a man's water needs at three activity levels in relation to the daily mean air temperature.

Arid Area Hazards

There are several hazards unique to desert survival. These include insects, snakes, thorned plants and cacti, contaminated water, sunburn, eye irritation, climatic stress, and poor personal habits.

Insects of almost every type abound in the desert. Lice, mites, wasps, and flies, which are drawn to man as a source of moisture and food, are extremely unpleasant and may carry disease. Old buildings, ruins, and caves are favorite habitats of spiders, scorpions, centipedes, lice and mites. These areas, which provide protection from the elements, also attract other wildlife. Therefore, take extra care when staying in these areas. Wear gloves at all times in the desert. Do not place your hands anywhere without first looking to see what is there. Visually inspect an area before sitting or lying down. When you get up, shake out and inspect your boots and clothing.

Snakes are found in all arid areas. They inhabit ruins, native villages, garbage dumps, caves, and natural rock outcroppings that offer shade. Never go without boots or walk through these areas without carefully inspecting them for snakes. Pay attention to where you place your feet and hands. Most snakebites result from stepping on or handling snakes. Avoid them. Once you spot them, give them wide berth.

Water found in arid areas may be contaminated. You should always assume so and treat it properly. Drinking untreated water will result in dysentery, causing excessive water loss.

Sunburn results from overexposing your skin to the sun's rays. So keep your body completely clothed, including gloves on your hands and a scarf around your neck. Use sunburn cream liberally on any exposed areas of skin. Sun poisoning equals nausea and dehydration. In addition, burns may become infected, causing more problems.

Remember that—

- There is as much danger of sunburn on cloudy days as on sunny days, especially at high altitudes.
- Sunburn ointment does not give complete protection against excessive exposure.
- Sunbathing or dozing in the desert sun can be fatal.

The glare on the sand causes eyestrain, and wind-blown, fine sand particles can irritate the eyes and cause inflamation. Wear goggles and use eye ointments to protect your eyes.

The combination of wind and sand or dust can cause your lips and other exposed skin to chap. Use chapstick and skin ointments to prevent or overcome this problem.

The sudden and extreme temperature shifts in arid areas can cause chest colds. Wear warm clothes at night to prevent chills.

Desert environments can cause stress to the body and mind. Some stress-rendering factors are:

- Extreme heat during the day with sudden temperature drops in the evening.
- Constantly blowing, sand-laden winds.
- Extensive barren areas, which bring on depression.

Rest is *essential* in this environment: You need 20 minutes of rest for each hour in the heat and you need 6 hours of sleep each day.

Proper disposal of human waste is essential. Bury feces and cover urine to prevent attracting flies. If possible, wash hands after defecating or urinating and before each meal. Clean eating and cooking utensils. It is important that you follow good sanitation procedures to lessen the danger of gastrointestinal disorders, which lead to excessive moisture loss.

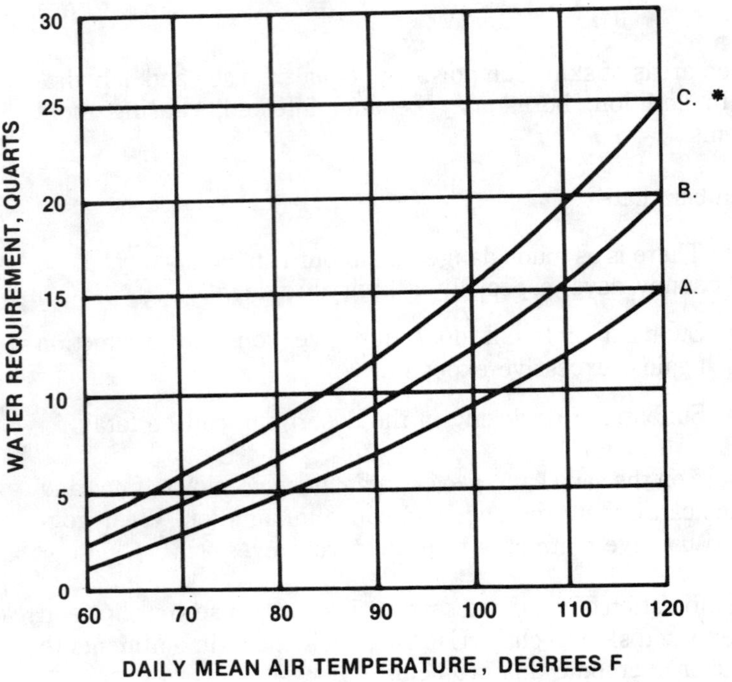

This graph shows water needs, in quarts per day, for men at three activity levels in relation to the daily mean air temperature. For example, if one is doing 8 hours of hard work in the sun (curve C) when the average temperature for the day is 100° F (horizontal scale), his water requirement for the day will be approximately 15 quarts (vertical scale).

From Technical Report E-P118, Southwest Asia: Environment and Its Relationship to Military Activities, July 1959.

*** C. HARD WORK IN SUN (creeping and crawling with equipment on)**

B. MODERATE WORK IN SUN (cleaning weapons and equipment)

A. REST IN SHADE

Figure 13-2. Daily water requirements for three levels of activity.

Chapter 14

Tropical Survival

You may think of the tropics as a jungle where a person must hack out a path and faces danger at every step. This is a misconception. Much of the land in the tropics is cultivated. Only a small portion is jungle (rain forest), and the greatest danger comes from insects and steep terrain. Much of the land is covered with secondary growth that has grown in the deserted farm fields.

The tropics are those areas of the world that lie between 23 1/2° north and 23 1/2° south of the equator. In the tropics are rain forests, semievergreen seasonal forests, tropical scrub and thorn forests, and tropical savannas.

You can enhance your chances of survival in these areas by—

- *Knowing how to perform the field skills covered in this manual.*
- *Having the ability to improvise.*
- *Being able to apply intelligently the principles presented in this manual.*
- *Learning what types of climate, terrain, and plant and animal life prevail in these areas.*
- *Learning about the hazards that exist and how to overcome these hazards.*
- *Maintaining your will to survive (Chapter 1).*

Rain Forests

Locations. Tropical rain forests are found in America, Asia, and Africa, and almost all are bisected by the equator (figure F-2). These rain forests are sometimes interspersed with mountain ranges and plateaus. They may even have some semidesert areas.

Characteristics. Following are some characteristics of tropical rain forests:

- Yearly rainfall is high (100 inches or more) and is more or less equally distributed throughout the year.
- There are five stories of vegetation (figure 14-1).
- Most of the trees are evergreen and many are large in girth (10 feet in diameter) with thick leathery leaves.
- Tree bark is generally thin, green, and smooth and usually lacks fissures.
- Vines and air plants abound.
- Herbs, grasses, and bushes are rare in the understory.
- There is generally uniformity in foliage in a well developed rain forest.

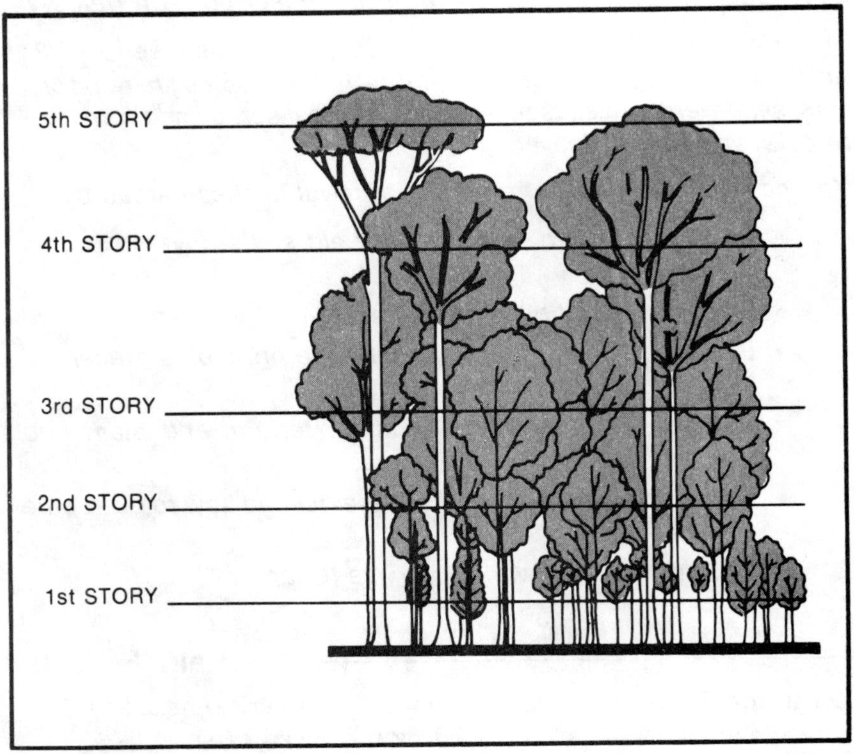

Figure 14-1. Five stories of tropical rain forest vegetation.

Trees are the principal vegetation in the rain forest. The tallest trees average 150 to 180 feet in height, although occasionally you may find one that reaches 300 feet. The trunks are usually straight and slender with the branches near the top. Many trees have a base of flangelike outgrowths. The majority of mature tropical trees have large, leathery, dark green leaves that resemble laurel leaves in shape, size, and texture.

The majority of other plants are woody with dimensions comparable to the trees. These include the vines and air plants, which grow on tree trunks and branches, and the bamboos, which are really giant grasses. In some areas bamboo grows 20 to 80 feet high and forms an impenetrable thicket.

There are many climbing plants, the majority having thick, long, woody stems. Some vines cling closely to the trees that support them, but most rise to the forest canopy like cables or hang down in loops or festoons.

The undergrowth consists of woody plants—seedlings and sapling trees, shrubs, and young woody climbers. Within old undisturbed forests, the undergrowth is usually thin enough to allow you to walk any direction. However, on river banks or in clearings where much light reaches the ground, the growth is dense and often impenetrable.

It is rare to find stands of just one kind or a few kinds of trees in the rain forest. Plants are often scattered, so you may have to search to find those with edible parts.

With midsummerlike weather throughout the year, vegetation is much the same year round. There are periods, however, when more species bloom and more young leaves are produced than at any other time. But for the most part, plant growth and reproduction are continuous and some flowers can be found at any time.

The general aspect of the rain forest is monotonous since large and strikingly colored flowers are uncommon. Most of the trees and shrubs have inconspicuous flowers, often greenish or whitish.

Food Sources. Foraging for food in the center of a virgin rain forest is difficult as fruits and nuts on trees are generally too high to reach and other types of plant food are scarce. At the edge of the rain forest, in clearings, and in areas around abandoned buildings, however, edible plants abound. Some food plants found in the rain forest are listed below:

bamboo	tropical almond	sugar palm
banana	bael fruit	Batoko plum
cattail	goa bean	pokeweed
nutsedge	cashew nut	air potato
water lily	water chestnut	rose apple
coconut palm	wild fig	tamarind
nipa palm	wild grape	taro
rice	water lettuce	ti plant
sweetsop	buri palm	tree fern
sugarcane		tropical yam

Many animals living in the jungle are good food sources. Look on the ground for hedgehogs, porcupines, anteaters, mice, wild pigs, deer, and wild cattle. Look in the trees for bats, squirrels, rats, and monkeys. Use the techniques described in Chapter 7 to capture and prepare the game for eating.

If you are near a stream or river, catch fish or other water wildlife (Chapter 7). Keep in mind that fish spoil quickly in the tropics; also, that mussels in tropical zones are poisonous during the summer.

Jungle Travel. Usually the key to jungle survival is travel. A rescue unit may be unable to pick you up immediately after contact, and you may have to travel many miles to reach a suitable pickup point.

Before you decide to move from your location, however, you should consider—

- Your chances of being found and recovered from your present location.
- Your physical condition.
- The availability of food and water. These usually abound in most jungles.
- The navigation equipment you have and your ability to use it.
- The personal equipment and clothing you have.

The most useful aids for jungle travel are:

- A machete to cut your way, to obtain food, and to make a raft.
- A compass to maintain direction.
- Medications to treat fever and infection.
- Stout shoes to protect your feet and to make walking easier.
- A hammock to lessen the time needed to prepare a bed above the ground.
- Mosquito netting to protect you from insects.

In seeking a route out of the jungle, you should seek the one that is *the safest* and offers the least resistance. Your situation, the weather conditions, and the type of terrain are major factors you must consider in selecting a route.

Often the route that offers the least resistance is a waterway. You should avoid cross-country travel if possible. Find a stream and travel downstream to a larger body of water. Set up camp, prepare signals, and wait for search aircraft. If you fail to make contact within a week, travel on downstream and set up another camp.

Following a stream requires fording, detouring, and penetrating thick vegetation. In mountainous country, a stream may tend to wind, the vegetation may be dense, observation points are rare, and swamps are common. Nevertheless, following a stream offers some advantages in strange country: It gives a definite course that will likely lead to habitations. It is a source of food and water, and possibly a means to travel by boat or raft.

If you are near a ridge, you will probably find travel easier on the ridge than in the valley. A ridge usually has less vegetation and less streams and swamps to cross than a valley. It can serve as a guide and afford observation points from which you can pick out landmarks. In addition, ridges generally have game trails along their tops. Take care, however, to conceal yourself if there are enemy personnel in the area.

Plan each day's travel so you have enough time and energy left to set up a secure and satisfactory campsite. Be sure to get enough rest and sleep before you travel on. How fast you can travel will depend on—

- Climatic conditions (temperature, sun, wind, and rain).
- Your physical condition and that of your companions if any.
- Terrain (angle of slope and type of footing).
- The location, characteristics, and number of enemy personnel in the area and whether they know your location.
- The amount of equipment you are carrying (carry only what you need).
- Your food requirements (if possible, hunt and gather food while traveling).

Check the weather signs before you decide to travel. Cloud formations are foretellers of weather.

Move through the jungle *in daylight only* unless your situation prohibits. Avoid thickets and swamps. Move in one direction, but not in a straight line. Turn your shoulders, shift your hips, and bend your body to work your way along. Shorten or lengthen your stride and slow or speed up your pace to adjust to the type and density of the vegetation.

Move slowly and steadily through dense vegetation, but stop periodically to listen and take your bearings. Noise carries a long distance in the jungle.

Use your bayonet or machete to cut through dense vegetation. An upward cutting stroke makes less noise than a downward stroke. To avoid leaving a distinct trail, cut only enough vegetation to get through.

Use a stick or staff to part the vegetation to reduce the possibility of dislodging fire ants.

Do not grab brush or vines to help you along on slopes or over obstacles. They may have irritating spines or sharp thorns; also, they may not hold your weight.

Do not climb over logs if you can walk around them. By walking around them you save your strength and are less likely to be injured.

Disregard the pattern of trees and bushes directly in front of you. Focus your eyes *beyond* your immediate front. Rather than looking *at* the jungle, look *through* it. Stoop occasionally and look along the jungle floor. By developing "jungle eyes" you will avoid scratches, bruises, and loss of direction. You will also gain confidence.

Many jungle animals follow well established game trails. These trails may wind and crisscross, but they often lead to water or clearings. Before traveling on these trails, check for enemy forces and animals that can harm you. Enemy forces or native people can easily watch and have ambush sites on trails, so avoid trails unless you absolutely have no other choice for travel.

If you use the trails, check your bearings frequently to make sure the trails go in the direction you want to go. Watch for disturbed places—there may be a pitfall or trap.

Do not follow a trail that has an obvious barrier such as a rope or a grass mat. The trail may lead to an animal trap.

Campsite. In the tropics twilight generally lasts less than 30 minutes and darkness sets in early, so pick your campsite and set up camp before sunset.

In selecting your campsite—

- Do not camp too near a stream or pond, especially during the rainy season. Heavy rainfall, locally or upstream, can cause flashflooding without warning.
- Do not camp under dead trees or trees with dead limbs that might fall on you.
- Do not camp on game trails or near waterholes. The traffic may be heavy.

Cut away a great deal of underbrush around your campsite to give you room to move around and to allow your fire to ventilate. This will minimize insects, eliminate hiding places for snakes, and make you more visible to air searches.

—— Semievergreen Seasonal and Monsoon Forests

Areas of the world having tropical semievergreen and monsoon forests are—

- South America. Portions of Columbia and Venezuela and the Amazon basin.

- Africa. Portions of southeast coastal Kenya, Tanzania, and Mozambique.
- Asia. Northeastern India, much of Burma, Thailand, Indochina, Java, and parts of other Indonesian Islands.

The characteristics of the American and African semievergreen seasonal forests correspond with those of the Asian monsoon forests. These characteristics are as follows:

- There are two stories of tree strata. The trees in the upper story average 60 to 80 feet in height; those in the lower story average 20 to 45 feet.
- The trees average 2 feet in diameter.
- There is a seasonal drought during which leaves fall.

Except for the sago, nipa, and coconut palms, the same edible plants grow in these areas as in the tropical rain forests.

Tropical Scrub and Thorn Forests

Areas of the world having tropical scrub and thorn forests are—

- America. West coast of Mexico, Yucatan, Venezuela, and Brazil.
- Africa. Coastal northwest Africa and south central Africa—Angola, Zambia, Zimbabwe, Tanzania, and Malawi.
- Asia. Turkestan and India.

The main characteristics of these forests are—

- There is a definite dry season and a wet season that varies in length from year to year. The rains are mainly downpours that come with thunderstorms.
- Trees are leafless during the dry season and average 20 to 30 feet in height. There is tangled undergrowth in places.
- The ground is bare except for a few tufted plants that grow in bunches; grasses are rare.
- Predominate plants have thorns.
- Fires occur at intervals.

During the dry season in the tropical scrub and thorn forest areas, a person will have difficulty obtaining food plants. The chief kinds of food come from the following plant parts: tubers, rootstalks, bulbs, corms, pith, gums and resins, nuts, seeds, and grains.

During the wet season, plant food is more abundant. You should look for the plants listed below (see Chapter 18 for descriptions).

acacia, sweet	wild chicory	prickly pear
agave	wild fig	St. Johns bread
almond	juniper	tamarind
baobab	sea orach	water lily
wild caper	cashew nut	tropical yam

Tropical Savanna

Areas of the world having savannas are—

- South America. Parts of Venezuela, Brazil, and the Guianas.
- Africa. Southern Sahara (north central Africa and southern Sudan), northern Gold Coast, most of Nigeria, northeastern Zaire, northern Uganda, western Kenya, part of Malawi, part of Tanzania, Zimbabwe, Mozambique, and western Madagascar.

Some characteristics of these savannas are—

- They lie wholly within the tropical zone in South America and Africa.
- They look like broad, grassy meadows with trees spaced at wide intervals.
- The grasses are bunch grasses that often exceed 6 feet in height. There is a definite space between each bunch.
- The scattered trees usually appear stunted and gnarled like apple trees.
- The soil is frequently red.

For the most part, the savannas of South America have a long dry season and, comparatively, a short wet season. Both high and short grasses grow on these savannas, and during the rainy season, bright colored flowers appear between the grass bunches. The grains of these grasses and the underground parts of the seasonal plants that appear with and following the rains are the primary plant food sources.

There are two types of savannas in Africa: the high grass and the bunch grass.

The high grass savanna, which surrounds the tropical rain forest, contains coarse grasses that grow 5 to 15 feet high. Unless the natives burn the grass during the dry season, the savanna becomes almost impenetrable.

The bunch grass savanna comprises the greatest part of the African savannas. The grasses average 3 feet in height.

Both dwarf and large trees grow on the savannas, the largest being the monkeybread or baobab tree.

Some edible plants that are found on the savannas of South America and Africa are listed below:

amaranth	wild crabapple	purslane
wild apple	wild fig	sorrel
baobab	water plantain	tamarind
wild chicory	nutsedge	water lily

Dangers in the Tropics

Although some poisonous snakes and large animals live in the tropics, they are only a small danger. The greatest danger is from insects that transmit diseases or have poisonous bites or stings. Infection in a wound, even as small as a scratch, is also a great danger.

Mosquitoes, ticks, fleas, mites, leeches, spiders, scorpions, centipedes, chiggers, wasps, wild bees, and ants live in the moist tropics. Take the following steps to prevent bites or stings by these insects:

- Avoid areas where they may be prevalent.
- Use standard issue insect repellent, if available, on all

exposed areas of skin and on all clothing openings.

- Wear all clothing, especially at night.
- Tuck pant legs into boots, roll down and button sleeves, and button collars.
- Wear gloves and a mosquito headnet if available.
- Camp away from swamps.
- Sleep under mosquito netting if you have it; otherwise, smear mud on your face.

Ticks and fleas are blood-sucking parasites that attach themselves to warm-blooded animals. Some are carriers of infectious diseases. Ticks often abound in grassy places. They will get on your clothing and body. Brush ticks off your clothing. Check your skin for ticks at least once a day; fleck them off if possible. If they have attached themselves to your skin, follow the procedure on page 3-20.

Fleas are common in dry, dusty buildings. The females will burrow under your toenails or into your skin to lay their eggs. Remove them with a sterilized knife. Keep the cut clean. In India and Southern China, rats carry fleas that transmit bubonic plague. Finding dead rats usually means a plague epidemic in the rat population. Fleas may also transmit typhus fever.

In many parts of the tropics, especially Malaya and Indonesia, rats carry other parasites that cause jaundice and other fevers.

In many Far East tropical areas, tiny red mites carry a type of typhus fever. They live in the soil, burrowed a few inches into the ground, and are common in tall grass, in cutover jungle, and on stream banks. Do not lie or sit on the ground. Clear your camping ground and burn it off. Treat your clothing with insect repellent.

Leeches are most common in wet underbrush. You may pick them up from plants, from the ground, or in water. They will get through your shoe eyelets or over your shoe tops.

Spiders, centipedes, and scorpions are often abundant. A few spiders have poisonous bites. The large spiders called tarantulas rarely bite but are covered with short hard hairs that will irritate your skin if you touch them. Centipedes pinch and inject venom through leg-like fangs at the base of their head. The pinch is similar to a bad wasp sting. A scorpion has poison glands in its tail. Although some species do have a fatal sting, the sting of most species causes severe pain and sickness. Scorpions hide beneath stones and the loose bark of dead trees. They also often hide in shoes left on the ground during the night. Always shake out your socks, shoes, and clothing before putting them on. Inspect your bed for pests. A sting from any of them can cause swelling and pain.

Chiggers, wasps, wild bees, and ants are pests that may harm you. Many biting ants live in the branches and foliage of tropical trees. Hanging plants attached to mangrove branches are almost always inhabited by biting ants. Do not camp near an ant hill or an ant trail.

Never walk barefoot. Your shoes guard against crawling mites, ticks, and cuts, which can become infected.

Be alert for crocodiles or alligators in any tropical waters. They prefer to lie on banks or float like logs with just their eyes above the water. Take great care in fording deep streams, in bathing, and in getting near any water body, especially where crocodiles or alligators are evident. Avoid them at all times. If you must get in the water, do not thrash about as this will attract them.

Rivers in northeastern South America are infested with piranha fish. These small fish attack in schools and can devour a 300-pound hog in a few minutes.

In saltwater estuaries, bays, or lagoons, sharks known to attack man may swim close to shore. Many sharks in shallow water of tropical seas have attacked. Sharks longer than 4 feet are potentially dangerous. Not all sharks show fins above the water (Chapter 16). Barracudas have also been known to attack in waters that are murky or clouded.

The Portuguese man-of-war is another warm saltwater hazard. These jellyfish-like creatures have stinging tentacles that may be as long as 50 feet. Their sting is extremely painful and may even disable a swimmer.

Pacific islands have fairly regular diurnal tides. However, storms in the vicinity may interrupt the normal ebb and flow with prolonged periods of high and low tides. During any change of tide, a great volume of water passes through atoll inlets. Do not attempt to swim or raft across the deeper channels until the tide reaches a lull at peak or ebb. Do not get caught far out on a reef by incoming tides. These tides sweep over a reef with a current that makes walking difficult and swimming dangerous.

Surf is not dangerous unless you are in a weakened condition or unless storms have built the wave action above the normal 8-foot height. Waves do not break until they are almost on the reef, and they move shoreward in a definite cycle. If you must swim, take advantage of the lull between series of large waves to get through the surf. Head *into* the waves. If a large wave is ready to break in front of you in shallow water, dive, grab hold of a rock and hang on until the crest of the wave has passed. Let the declining force of the wave carry you shoreward.

In Chapter 7 you read about edible seafood that you can catch. But you should also be aware of some of the dangers on tropical seashores.

The flesh of many species of reef fish (page 7-18) contains toxins that are poisonous. Some fish that are considered edible, such as red snapper and barracuda, are poisonous when taken from atolls and reefs. Others have spines, stingers, or "teeth" that inject toxins into unwary persons.

Do not walk barefooted on coral reefs. Coral, dead or alive, can cut your feet to ribbons. Fine needles of lime or silica from seemingly harmless sponges and sea urchins can get in your skin and fester. The almost invisible stonefish will not move from your path. Its poisonous spines will cause you agony and possible death. Treat as for snakebite (page 3-22).

In tropical waters, use a stick to probe dark holes. Do not use your hands. When walking over muddy or sandy bottoms of rivers and seashores, do not step freely—slide your feet along the bottom to avoid stepping on stingrays or other sharp-spined animals.

Cone snails and long, slender, pointed terebra snails have a toxic sting. They live under rocks, in crevices of coral reefs, and along rocky shores and protected bays. Avoid handling all cone shells.

Handle the big conches with caution. These snails have razor-sharp trapdoors, which they may suddenly jab out, puncturing your skin as they try to get away.

Do not use your hands to gather large abalones and clams. Pry or wedge them loose with a stick or some such device. They will hold you if they clamp down on your fingers.

If you must cross deeper portions of a reef, check the reef edge for sharks, barracudas, and moray eels. Moray eels hide in dark holes among the reefs and are vicious and aggressive when disturbed.

Chapter 15

Arctic and Subarctic Survival

When you think of the arctic and subarctic regions, you probably think of extremely cold weather. And these regions do have long periods of extremely cold weather. However, in terms of temperature, these regions are best defined as follows:

Arctic—Those regions where the mean temperature of the warmest month of the year does not exceed 50° F (10° C).

Subarctic—Those regions where the mean temperature of the warmest four months of the year does not exceed 50° F (10° C).

Mountainous and high-elevation areas of temperate regions also have extremely cold weather, and cold weather survival information presented here also applies to these regions.

Cold Environment Characteristics

Cold weather is a dynamic force and can become a formidable adversary even though you recognize its hazards and take advantage of its peculiar characteristics. Ignoring or underestimating its force can result in death.

Windchill increases the hazards in cold regions. This is the effect of moving air on exposed flesh. For instance, with a 15-knot wind and a 15^0 F temperature, the equivalent chill temperature is -10^0 F. Looking at figure 15-1, you can see that the greater the wind speed, the lower the equivalent chill temperature.

Changes in weather and temperature occur rapidly in cold environments. These changes can affect your speed and ease of travel. For instance, rainfall, snowfall, or a rise in temperature may make traveling impossible or extremely dangerous over the type of terrain that you crossed easily and quickly the preceding day.

Obtaining your basic needs—food, water, and shelter—in a cold environment is more difficult than in a warm environment. And even if you have these basic needs, you must also have adequate protective clothing and the will to live.

In a cold environment, as in any environment, your first priority is taking care of any injuries or sickness you or your companions have.

Health Hazards

When you are healthy, your inner core temperature (torso temperature) remains almost constant at 98.6° F. Since your limbs and head have less protective body tissue than your torso, their temperature varies and may not reach core temperature.

Your body has a control system that enables it to react in an effort to maintain a temperature balance. There are three main factors that affect this temperature balance: heat production, heat loss, and evaporation. The rate of heat production is governed by the difference between the body's core temperature and the environment's temperature. Your body can get rid of heat better than it can produce it. Sweating helps to control the heat balance. Maximum sweating will get rid of heat about as fast as maximum exertion produces it.

Shivering causes the body to produce heat. But it also causes fatigue, which in turn leads to a drop in body temperature. Air movement around your body affects heat loss. It has been calculated that a naked man exposed to still, cold air can maintain a heat balance at about freezing point if he shivers as hard as he can. But he can't shiver forever.

It has also been calculated that a man at rest and wearing maximum arctic clothing in a cold environment can keep his heat balance well above freezing. To withstand really cold conditions for any length of time, however, he will have to become active or shiver.

Following are cold injuries that can occur:

Hypothermia. This occurs when a person's body temperature drops to between 95° and 77° F. It can even occur when air temperatures are above freezing. To prevent hypothermia, avoid

COOLING POWER OF WIND EXPRESSED AS "EQUIVALENT CHILL TEMPERATURE"																						
WIND SPEED		TEMPERATURE (°F)																				
CALM	CALM	40	35	30	25	20	15	10	5	0	-5	-10	-15	-20	-25	-30	-35	-40	-45	-50	-55	-60
KNOTS	MPH	EQUIVALENT CHILL TEMPERATURE																				
3-6	5	35	30	25	20	15	10	5	0	-5	-10	-15	-20	-25	-30	-35	-40	-45	-50	-55	-65	-70
7-10	10	30	20	15	10	5	0	-10	-15	-20	-25	-35	-40	-45	-50	-60	-65	-70	-75	-80	-90	-95
11-15	15	25	15	10	0	-5	-10	20	-25	-30	-40	-45	-50	-60	-65	-70	-80	-85	-90	-100	-105	-110
16-19	20	20	10	5	0	-10	-15	-25	-30	-35	-45	-50	-60	-65	-75	-80	-85	-95	-100	-110	-115	-120
20-23	25	15	10	0	-5	-15	-20	-30	-35	-45	-50	-60	-65	-75	-80	-90	-95	-105	-110	-120	-125	-135
24-28	30	10	5	0	-10	-20	-25	-30	-40	-50	-55	-65	-70	-80	-85	-95	-100	-110	-115	-125	-130	-140
29-32	35	10	5	-5	-10	-20	-30	-35	-40	-50	-60	-65	-75	-80	-90	-100	-105	-115	-120	-130	-135	-145
33-36	40	10	0	-5	-15	-20	-30	-35	-45	-55	-60	-70	-75	-85	-95	-100	-110	-115	-125	-130	-140	-150
WINDS ABOVE 40 HAVE LITTLE ADDITIONAL EFFECTS.		LITTLE DANGER					INCREASING DANGER (Flesh may freeze within 1 minute)						GREAT DANGER (Flesh may freeze within 30 seconds)									
		DANGER OF FREEZING EXPOSED FLESH FOR PROPERLY CLOTHED PERSONS																				

Figure 15-1. Chill index.

actions that can cause rapid or uncontrolled loss of body heat.

Symptoms of hypothermia are sluggish movement, reduced coordination, and impaired judgment. If the victim's core temperature drops below 77° F, death is almost certain.

To treat hypothermia, rewarm the entire body. If there are means available, rewarm the person by first immersing the trunk area only in warm water 100° to 110° F.

CAUTION: *Rewarming of the total body in a warm water bath should be done only in a hospital environment because of increased risk of cardiac arrest and rewarming shock.*

One of the quickest ways to get heat to the inner core is to give warm water enemas. Another method is to wrap the victim in a warmed sleeping bag with another person who is already warm; both should be naked. If the person is conscious, give him hot sweetened fluids. One of the best sources of calories is honey or dextrose; if unavailable, use sugar, cocoa, or a similar soluble sweetener.

CAUTION: *Do not force an unconscious person to drink.*

There are two dangers in treating hypothermia: too rapid rewarming and "after drop." Too rapid warming can cause the victim to have circulatory problems, resulting in heart failure. After drop is the sharp temperature drop that occurs when the victim is taken from the warm water. This is probably caused by previously stagnant limb blood being returned to the core (inner torso) area as recirculation occurs. Concentrating on warming the core area without stimulating peripheral circulation will lessen the effects of after drop. Immersing the torso in a hot bath, if possible, is the best treatment.

Frostbite. This is an injury resulting from frozen tissues. Light frostbite involves only the skin, which takes on a dull whitish pallor. Deep frostbite extends to a depth below the skin. The tissues become solid and immovable. Your feet, hands, and exposed facial areas are particularly vulnerable to frostbite.

The best prevention for frostbite when you are with others is to use the buddy system. Check your buddy's face often and make sure that he checks yours. If you are alone, periodically cover your nose and lower part of your face with your mittened hand.

A loss of feeling in your hands and feet is an indication of frostbite. If you have had only a short period of time without feeling, the frostbite is probably light. Otherwise, assume the frostbite is deep.

Figure 15-2 lists some do's and don'ts regarding frostbite.

A deep frostbite injury once thawed and refrozen will cause more damage than a nonmedically trained person can handle.

Trench Foot and Immersion Foot. These conditions result from many hours or days of exposure to wet or damp conditions at a temperature just above freezing. The feet become cold, swollen, and have a waxy appearance. Walking becomes difficult and the feet feel heavy and numb. The nerve and muscles sustain the main damage, but gangrene can occur. In extreme cases the flesh dies and it may become necessary to have the foot or leg amputated. The best preventive is to keep the feet dry. Carry extra socks with you in a waterproof packet. Wet socks can be dried against the body. Wash your feet daily and put on dry socks.

Dehydration. In cold weather, bundled up in many layers of clothing, you may be unaware that you are losing body moisture. Your heavy clothing absorbs the moisture, which evaporates in the air. You must drink water to replace this loss of fluid. Your need for water is as great in a cold environment as it is in a warm environment (Chapter 3). One way to tell if you are becoming dehydrated is to check the color of your urine on snow. If your urine makes the snow dark yellow, you are becoming dehydrated and need to replace body fluids. If it makes the snow light yellow to no color, your body fluids have a more normal balance.

DO'S	DON'TS
Periodically check for frostbite. Rewarm light frostbite. Keep injured areas from refreezing.	Don't rub injury with snow. Don't drink alcoholic beverages. Don't smoke. Don't try to thaw out a deep frostbite injury if you are away from definitive medical care.

Figure 15-2. Frostbite do's and don'ts.

Cold Diuresis. This is an increased output of urine caused by exposure to cold. It also decreases the body fluids, which must be replaced.

Sunburn. Exposed skin can become sunburned even when the air temperature is below freezing. The sun's rays reflect at all angles from snow, ice, and water, hitting sensitive areas of skin: the lips, nostrils, and eyelids. More sunburn results at high altitudes than at low altitudes for the same time exposed to the sun. Apply sunburn cream or lip salve to your face when you are exposed to the sun's rays.

Snow Blindness. This is caused by the reflection of ultraviolet rays caused by the sun shining brightly on a snow-covered area. The symptoms of snow blindness are the feeling of grit in the eyes, pain in and over the eyes that increases with eyeball movement, eyes watering and becoming red, and a headache, which intensifies with continued exposure to light. Prolonged exposure to these rays can result in permanent eye damage. To treat snow blindness, bandage the eyes until the symptoms disappear.

You can prevent snow blindness by wearing your sunglasses. If you don't have sunglasses, improvise: Cut slits in a piece of cardboard, thin wood, tree bark, or other available material (figure 15-3). Putting soot under your eyes will help reduce shine and glare.

Figure 15-3. Improvised sunglasses.

Constipation. Putting off relieving yourself because of the cold, eating dehydrated foods, drinking too small amount of liquid, and irregular eating habits can cause you to become constipated. Although not disabling, constipation can cause some discomfort. Increase your fluid intake to at least two quarts per day and eat fruits, if available, and other foods that will loosen the stools.

Hygiene

Although washing yourself may be impractical and uncomfortable in a cold environment, you must do so. Washing helps to prevent skin rashes that can develop into more serious problems.

In some situations, you may be able to take a snow bath. Take a handful of snow and wash your body where sweat and moisture accumulate, such as under the arms and between the legs, front and rear, and then wipe yourself dry. If you cannot bathe, periodically wipe yourself dry in these areas. If possible, wash your feet daily and put on clean dry socks. Change your underwear at least twice a week. If you are unable to wash your underwear, take it off, shake it, and let it air out for an hour or two.

If you are with natives or are using a shelter that has been used before, check your body and clothing each night for lice. If your clothing has become infested, use insecticide powder if you have any. Otherwise, hang your clothes in the cold, then beat and brush them. This will help get rid of the lice but not the eggs.

Clothing and Equipment

You must not only have enough clothing to protect you from the cold, you must also know how to get the most warmth from it. For instance, you should keep your head covered. You can lose a lot of body heat from an unprotected head, neck, wrists, and ankles.

There are four basic principles you should follow to keep warm. These are:

- Wear your clothing loose and in layers. Wearing too tight clothing and footgear restricts the circulation of the blood and invites cold injury. It also decreases the volume of air trapped between the layers, reducing its insulating value.

Several layers of lightweight clothing are better than one equally thick layer of clothing, because the layers have dead air space between them. The dead air space provides extra insulation. In addition, layers of clothing allow you to take off or add clothing layers to prevent excessive sweating or to increase warmth.

- Avoid overheating. When you get too hot, you sweat and your clothing absorbs the moisture. This affects your warmth in two ways: dampness decreases the insulating quality of clothing, and as sweat evaporates your body cools. Adjust your clothing so you do not sweat. You can do this by partially opening your parka or jacket, by removing an inner layer of clothing, by removing heavy mittens, or by throwing back your parka hood or changing to lighter head cover. The head and hands act as efficient heat dissipators when overheated.

- Keep clothing dry. In cold temperatures, your inner layers of clothing can become wet from sweat and your outer layer, if not water repellent, can become wet from snow and frost melted by body heat. Wear water repellent outer clothing, if available. It will shed most of the water collected from melting snow and frost. Before entering a heated shelter, brush off the snow and frost. Despite the precautions you take, there will be times when you cannot keep from getting wet. At such times, drying your clothing may become a major problem. On the march, hang your damp mittens and socks on your pack. Sometimes in freezing temperatures, the wind and sun will dry this clothing. Or you can put damp socks or mittens, unfolded, near your body so that your body heat can dry them. In bivouac, hang damp clothing inside the tent near the top, using drying lines or improvised racks. You may even be able to dry each item by holding it before an open fire. Dry leather items slowly. If no other means are available for drying your boots, put them between the sleeping bag shell and liner. Your body heat will help to dry the leather.

- Keep clothing clean. This is always important from the standpoint of sanitation and comfort. In winter, it is also important from the standpoint of warmth. Clothes matted with dirt and grease lose much of their insulation quality.

If the air pockets in clothing are crushed or filled up, heat can escape from the body more readily.

A heavy down-lined mummy bag is one of the most valuable pieces of survival gear in cold weather. Make sure the down remains *dry*. If wet, it loses a lot of its insulation value. If you do not have a sleeping bag, you can make one out of parachute cloth or similar material and natural dry material, such as leaves, pine needles, or moss. Place the dry material between two layers of parachute cloth.

Other important survival items are a knife; waterproof matches in a waterproof container, preferably one with a flint attached; a durable compass; map; watch; waterproof groundcloth and cover; flashlight; binoculars; dark glasses; fatty emergency foods; food-gathering gear; and signaling items. (See Chapter 2 for additional information on survival equipment.)

Shelter

The equipment you carry with you and your environment will determine the type of shelter you can build.

NOTE: *In extreme cold do not use metal, such as an aircraft fuselage, for shelter. The metal will conduct what little heat you can generate away from the shelter.*

The tree-pit shelter described in Chapter 8 may be the easiest to build and most effective for some situations. Figures 15-4, 15-5, and 15-6 show other shelters that you can make, depending on the materials and tools you have.

Shelters made from ice or snow usually require tools such as ice axes or saws. They also require a lot of time and energy to make.

Be sure to ventilate your shelter if you intend to build a fire in it.

Always make sure the entrance is clear of snow and free of obstacles that might prevent a fast exit.

Never sleep directly on the ground. Lay down some pine boughs, grass, or other insulating material to keep the ground from absorbing your body heat.

Never fall asleep without turning out your stove or lamp. Carbon monoxide is a great danger. It is colorless and odorless. It is freely generated by a yellow flame, so if you see a yellow flame, check your ventilation.

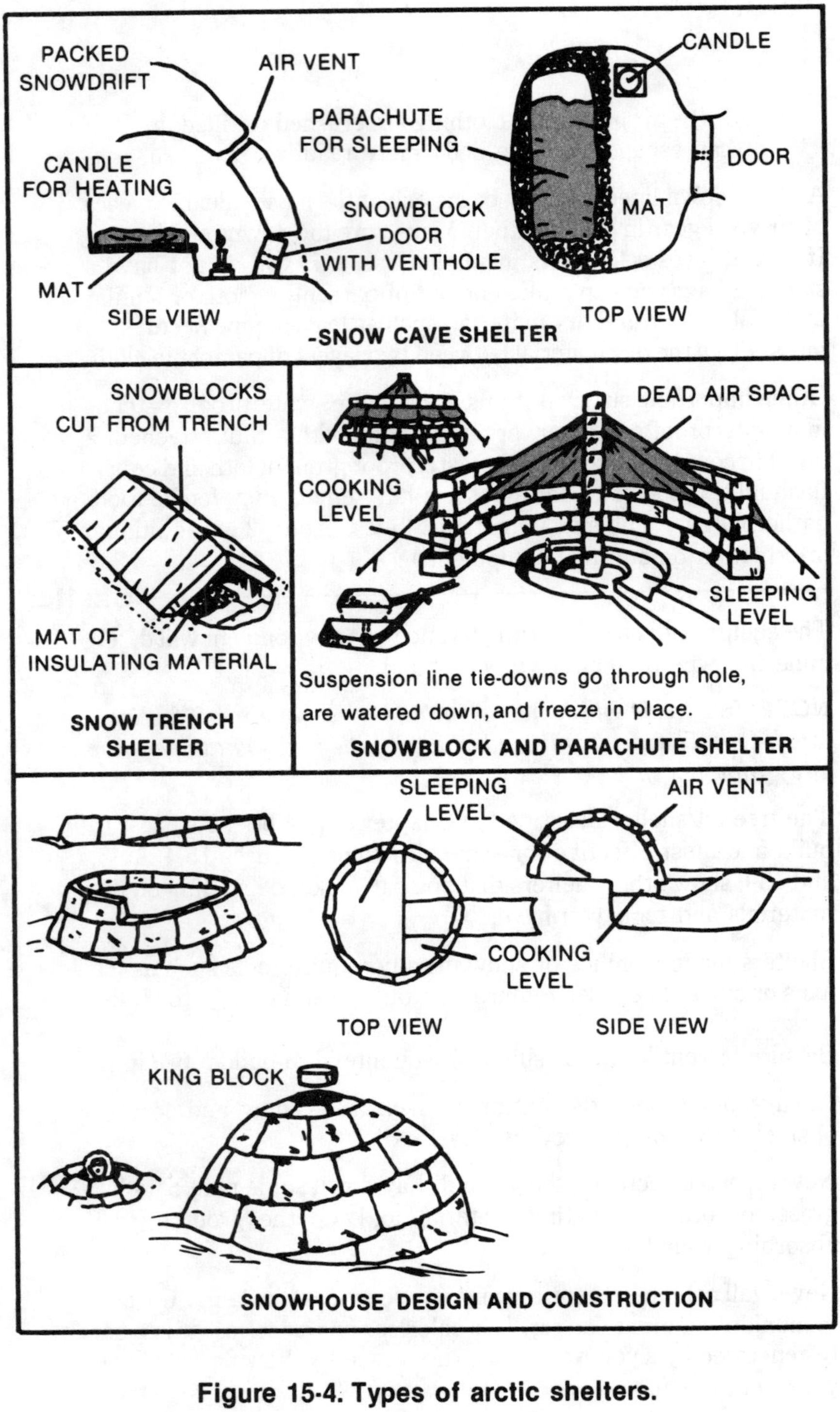

Figure 15-4. Types of arctic shelters.

Figure 15-5. Lean-to made from natural shelter.

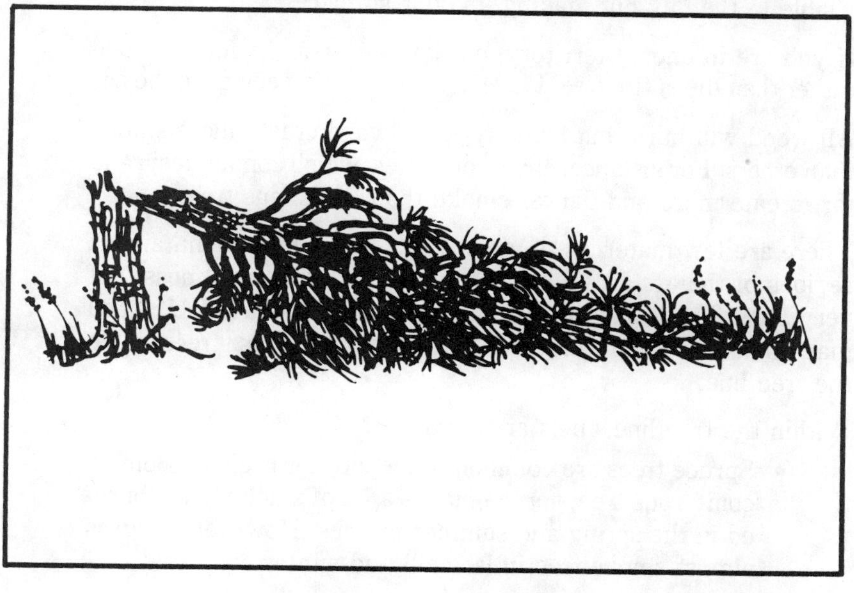

Figure 15-6. Fallen tree as shelter.

Carbon monoxide poisoning can result from a fire burning in an unventilated shelter. Even in a ventilated shelter, incomplete combustion can cause carbon monoxide poisoning. Usually there are no symptoms; unconsciousness and death can occur without warning. Sometimes, however, pressure at the temples, burning of the eyes, headache, pounding pulse, drowsiness, or nausea may occur. Get into fresh air at once if you have any of these symptoms.

Fire

Fire is especially important in cold weather. It not only provides a means for preparing food, but for getting warm and for melting snow or ice for water.

Use the techniques described in Chapter 9 for building and lighting your fire. If you are in enemy territory, however, keep in mind that the smoke, smell, and light from your fire may reveal your location. Light reflects from surrounding trees or rocks, making even indirect light a source of danger. Smoke tends to go straight up in cold, calm weather, making it a beacon during the day but helping to conceal the smell at night. In warmer weather, especially in a wooded area, smoke tends to hug the ground, making it less visible in the day but making its odor spread.

If you are in enemy territory, cut low tree boughs for firewood rather than the entire tree. Fallen trees are easily seen from the air.

All wood will burn, but some types of wood create more smoke than others. For instance, coniferous trees, which contain resin and tar, create more and darker smoke than deciduous trees.

There are few materials to use for fuel in the high mountainous regions of the arctic. You may find some grasses and moss, but very little. The lower the elevation, the more fuel available. You may find some scrub willow and small, stunted spruce trees above the tree line.

Within the tree line, fuels are abundant:

- Spruce trees are common in the interior regions. Being a coniferous tree, spruce makes a lot of smoke when burned in the spring and summer months. However, it burns almost smoke-free in late fall and winter.
- The tamarack tree is also coniferous. It is the only tree of the pine family that loses its needles in the fall. Without

its needles it looks like a dead spruce, but it has many knobby buds and cones on its bare branches. When burned, tamarack wood makes a lot of smoke and is therefore good for signaling purposes.

- Birch trees are deciduous and the wood burns hot and fast, much like it has been soaked with oil or kerosene. Most trees grow near streams and lakes, but occasionally a few will be found on higher ground and away from water.
- Willow and alder grow in arctic regions, normally in marsh areas or near lakes and streams. These woods burn hot and fast without much smoke.

Dried moss, grass, and scrub willow are other materials you can use for fuel. These are usually plentiful near streams in tundra (open, treeless plains) areas. By bundling or twisting grasses or other scrub vegetation to form a large, solid mass, you will have a slower burning, more productive fuel.

If there is fuel or oil available from a wrecked vehicle or downed aircraft, use it for fuel. Leave the fuel in the tank for storage, drawing on the supply only as you need it. Oil congeals in extreme cold temperature, so drain it from the vehicle or aircraft while still warm if there is no danger of explosion or fire. If you have no container, let the oil drain onto the snow or ice. Scoop up the fuel as you need it.

CAUTION: *Do not expose flesh to petroleum, oil, and lubricants (POL) in extreme cold temperatures. The liquid state of these products is deceptive in that it can cause frostbite.*

Some plastic spoons, helmet visors, visor housings, and foam rubber will ignite quickly from a burning match and will burn long enough to aid in starting a fire. For example, a plastic spoon will burn for about 10 minutes.

In cold weather regions, there are some hazards in the use of fires, whether for keeping warm or for cooking.

Fires have been known to burn underground resurfacing close by. Therefore, do not build a fire too close to a shelter.

In snow shelters, excessive heat will melt the insulating layer of snow.

A fire inside a shelter that lacks adequate ventilation can result in carbon monoxide poisoning.

A person trying to get warm or to dry clothes may become careless and burn or scorch his clothing and equipment.

A small fire about the size of a man's hand is ideal for an evader. It requires very little fuel, yet it generates considerable warmth and is hot enough to warm liquids.

A single candle provides enough heat to warm an enclosed shelter.

A small fire and some type of stove is best for cooking purposes. A hobo stove (figure 15-7) is particularly suited to the arctic. It is easy to make out of a tin can, and it conserves fuel.

A simple crane propped on a forked stick will hold a cooking container over a fire.

A bed of hot coals provides the best cooking heat. Coals from a crisscross fire will settle uniformly. This type fire is made by placing the firewood crisscross.

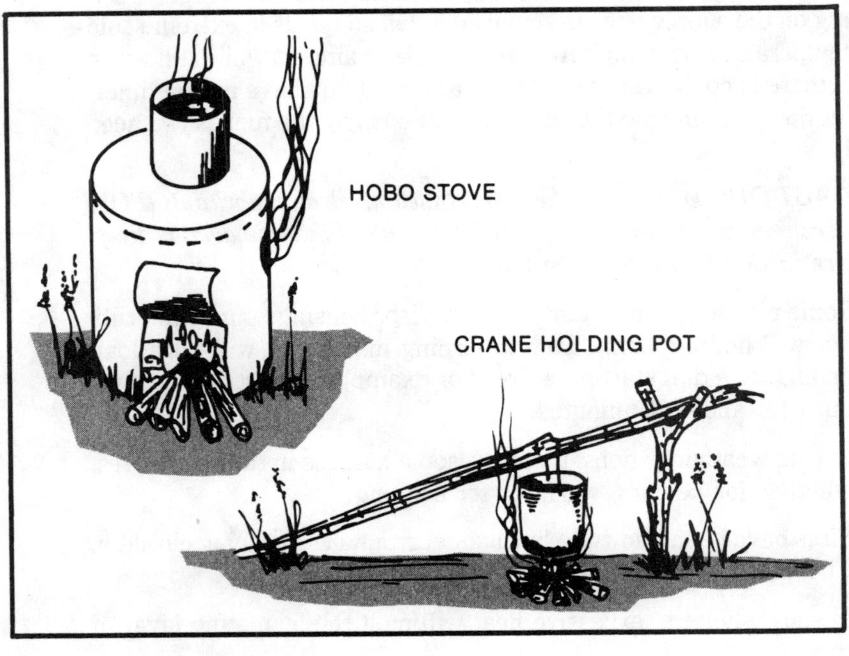

Figure 15-7. Cooking fire/stove.

Water

There are many sources of water in the arctic and subarctic. Your location and the season of the year will determine where and how you obtain water.

Water sources in arctic and subarctic regions are more sanitary than in other regions due to the climatic and environmental conditions. However, you should always purify the water before drinking it (Chapter 5).

During the summer months, the best natural sources of water are freshwater lakes, streams, ponds, rivers, and springs. Water from ponds or lakes may be slightly stagnant, but still usable. Running water in streams, rivers, and bubbling springs is usually fresh and suitable for drinking.

The brownish surface water found in tundra during the summer is a good source of water. However, you may have to filter the water before purifying it.

Freshwater ice and snow can be melted for water. Be sure they are completely melted before putting into your mouth. Trying to melt ice or snow in your mouth takes away body heat and may cause internal cold injuries.

You can use body heat to melt snow. Place the snow in a water bag and place the bag between your layers of clothing. This is a slow process, but you can use it on the move or when you have no fire.

NOTE: *Do not waste fuel to melt ice or snow when drinkable water is available from other sources.*

When ice is available, melt it rather than snow because one cup of ice yields more water than one cup of snow. Ice also takes less time to melt.

You can melt ice or snow in a water bag, ration can, or improvised container by placing the container near a fire. Begin with a small amount of ice or snow in the container and, as it turns to water, add more ice or snow.

Another way to melt ice or snow is by putting it in a bag made of porous material and suspending the bag near the fire. Place a container under the bag to catch the water.

During cold weather, avoid drinking a lot of liquid before going to bed. Crawling out of a warm sleeping bag at night to relieve yourself means less rest and more exposure to the cold.

Food Sources

There are several sources of food in the arctic and subarctic regions. The type of food—fish, animal, fowl, or plant—and the ease in obtaining it depend on the time of the year and your location.

Fish. During the summer months, fish and other waterlife are easily obtained from coastal waters, streams, rivers, and lakes. Use the techniques described in Chapter 7 to catch them.

The North Atlantic and North Pacific coastal waters are rich in seafood. You can easily find crawfish, snails, clams, oysters, and king crab. During low tide in areas where there is a great difference between the high-tide and low-tide water levels, you can easily find shellfish. Dig in the sand on the tidal flats. Look in tidal pools and on offshore reefs. In areas where there is a small difference between the high-tide and low-tide water levels, storm waves often wash shellfish onto the beaches.

The eggs of the spiny sea urchin that lives in the waters around the Aleutian Islands and southern Alaska are excellent food. Look for the sea urchins in tidal pools. Break the shell by placing it between two stones. The eggs are bright yellow in color.

Most northern fish and fish eggs are edible. Exceptions are the meat of the arctic shark and the eggs of the sculpins.

The bivalves, such as clams and mussels, are generally more palatable than spiral-shelled seafood, such as snails.

CAUTION: The black mussel, which is one of the most common mollusks of the far north, may be poisonous in any season. Toxins sometimes found in the mussel's tissue are as dangerous as strychnine.

The sea cucumber is another edible sea animal. Inside its body are five long white muscles that taste much like clam meat.

In early summer, smelt spawn in the beach surf. Sometimes you can scoop them up with your hands.

Kelp, the long ribbon-like seaweed, and other smaller seaweed that grow among offshore rocks are edible. You can often find herring eggs on the seaweed in midsummer.

Sea Ice Animals. Polar bears are found in practically all arctic coastal regions, but are rarely found inland. Avoid them if possible. They are the most dangerous of all bears. They are tireless, clever hunters with good sight and an extraordinary sense of smell. If you must kill one for food, approach it cautiously. Aim for the brain; a bullet elsewhere will rarely kill one. Always cook polar bear meat before eating it.

CAUTION: Do not eat the liver as it contains a dangerous concentration of vitamin A.

Earless seal meat is one of the best. You need considerable skill, however, to get close enough to an earless seal to kill it. In spring, seals often bask on the ice beside their breathing holes. They raise their heads about every 30 seconds, however, to look for their enemy, the polar bear.

To approach a seal, do as the Eskimos do: Stay downwind from it, cautiously moving closer while it sleeps. If it moves, stop and imitate its movements by lying flat on the ice, raising your head up and down, and wriggling your body slightly. Approach the seal with your body sideways to it and your arms close to your body so that you look as much like another seal as possible. The ice at the edge of the breathing hole is usually smooth and at an incline, so the least movement of the seal may cause it to slide into the water. Therefore, try to get within 25 to 50 yards of the seal and kill it instantly (aim for the brain). Try to reach the seal before it slips into the water. In winter a dead seal will usually float, but it is difficult to retrieve from the water.

Keep the seal blubber and skin from coming into contact with any scratch or broken skin you may have. You could get what is called "spekk-finger," which causes the hands to become badly swollen.

Keep in mind that where there are seals, there are usually polar bears, and polar bears have been known to stalk and kill seal hunters.

Bearded seal and walrus stay on floe ice. A seal is curious and you can sometimes attract it to gunshot range. Walrus are indolent, but extremely dangerous at close quarters. It is best to approach them by boat. Kill the walrus and seal on the ice rather than in the water so that you can get to the carcasses easily. Shoot walrus through the neck just below the head. You can sometimes get milk as well as meat from a dead walrus. You can milk a dead walrus, which will often yield as much as 16 quarts of milk. This same walrus, when milked an hour later, will give almost the same volume of milk.

CAUTION: *Do not eat liver of bearded seal. It, like the polar bear liver, contains a high concentration of vitamin A.*

Land Animals. Moose, caribou, mountain goats and sheep, musk ox, and bear are found in some arctic and subarctic regions.

Moose are more often found in heavy brush and may charge a person. In the winter you can sometimes spot moose by climbing a hill or tree and looking for the animal's "smoke" (condensed body vapor that rises like the smoke of a small fire).

Caribou migrate throughout northern arctic areas. They can be found in Alaska, northern Canada, western Greenland, and Siberia. In winter they feed on the tundra. In summer, they move close to the sea or into the high mountains.

In winter, mountain sheep descend to lower elevations and to valley feeding grounds. They are wary and hard to approach. By staying on higher ground than they are on and approaching them from their downwind side while they are eating, you can sometimes get close enough to shoot them.

Musk oxen are found in northern Greenland and the islands of the Canadian archipelago. Their tracks and droppings are similar to those of cattle. When alarmed, musk oxen group together. If approached, one or more bulls may charge.

Bears are dangerous, especially so if startled, wounded, or with young. If you find a large area of torn-up sod, it is likely that a bear has been digging for roots or ground squirrels. Do not shoot a bear unless you are sure you can kill it. Aim at the base of the ear, the neck, or just behind the shoulder. Bears hibernate during the winter.

Foxes, rabbits, ground squirrels, lemmings, and mice live on the tundra. Marmots live in the mountains, among rocks, usually near the edge of a meadow, or in deep soil. Ground squirrels and marmots hibernate during the winter.

Wolves are also found in arctic and subarctic regions.

Porcupine can be found in southern subarctic regions where there are trees. Porcupine feed on bark, so if you find tree limbs stripped bare, you are likely to find a porcupine in the area.

Ptarmigan, owls, and ravens are the only birds that remain in the arctic during the winter, and they are scarce north of the tree line. Ptarmigan and owl are as good for food as any game bird. Ravens are too thin to be worth the effort it takes to catch them. Ptarmigan, which change color to blend with their surroundings, are hard to spot. Rock ptarmigans travel in pairs and are easily approached. Willow ptarmigans live among willow clumps in bottom lands. They gather in large flocks and are easily snared. During the summer months all arctic birds have a 2- to 3-week moulting period during which they are flightless and easily caught. Use one of the techniques described in Chapter 7 to catch them.

Skin and butcher game (see Chapter 7) while it is still warm. If you do not have time to skin the game, at least remove its entrails, musk glands, and genitals before storing. If time allows, cut the meat into usable pieces and freeze each separately so you can use the pieces as needed. Leave the fat on all animals except seals. During the winter, game freezes quickly left in the open. During the summer, you can store it in ground ice holes.

Plants. Although treeless, tundra has a great variety of plants that grow during the warm months. All are small, however, when compared to plants in warmer climates. For instance, the arctic willow and birch are shrubs rather than trees.

Some food plants found in arctic and subarctic regions are listed below:

salmonberry
cranberry
crowberry
bilberry
bearberry
spadderdock
Eskimo potato
woolly lousewort
bistort

dandelion
marsh marigold
arctic willow
fireweed
coltsfoot
lichens

There are some plants growing in arctic and subarctic regions that are poisonous if eaten. Stick to plants that you know are edible. When in doubt, follow the Universal Edibility Test (page 6-4).

Travel

As a survivor or an evader in an arctic or subarctic region, you will face many obstacles. Your location and the time of the year will determine the types of obstacles and the inherent dangers. You should—

- Avoid traveling during a blizzard.
- Take care when crossing thin ice. Distribute your weight by lying flat and crawling.
- Cross streams when the water level is lowest. Normal freezing and thawing action may cause a stream level to vary as much as 6½ to 8 feet (2 to 2½ meters) per day. This may occur any time during the day, depending on the distance from a glacier, the temperature, and the terrain. You should also consider this variation in water level when selecting a campsite near a stream.
- Consider the clear arctic air. It makes estimating distance difficult. Distances are more frequently underestimated than overestimated.
- Avoid traveling in "whiteout" conditions. The lack of contrasting colors makes it impossible to judge the nature of the terrain.
- Always cross a snow bridge at right angles to the obstacle it crosses. Find the strongest part of the bridge by poking ahead with a pole or ice axe. Distribute your weight by crawling or by wearing snowshoes or skis.

- Make camp early so that you have plenty of time to build a shelter before dark.
- Consider rivers, frozen or unfrozen, as avenues of travel. Frozen rivers are frequently clear of loose snow, making travel on them easier than on the land.
- Use snowshoes if you are traveling over snow-covered terrain. Snow 12 or more inches deep makes traveling difficult and can lead to trenchfoot or frostbite if footwear becomes wet. If you do not have snowshoes, make a pair using willow, strips of cloth, leather, or other suitable material.

It is almost impossible to travel in deep snow without snowshoes or skis. And traveling by foot leaves a well marked trail for any pursuers to follow. If you must travel in deep snow, avoid snow-covered streams. The snow, which acts as an insulator, may have prevented ice from forming over the water. In hilly terrain, avoid areas where avalanches appear possible. Travel in the early morning in areas where there is danger of avalanches. On ridges, snow accumulates on the lee side in overhanging piles, called cornices. These often extend far out from the ridge and may break loose if stepped on.

Weather Signs

There are a number of good indicators of climatic changes.

Wind. You can determine wind direction by dropping a few leaves or grass or by watching the tops of trees. Once you determine the wind direction, you can predict the type of weather that is imminent. Rapidly shifting winds indicate an unsettled atmosphere and a likely change in the weather.

Clouds. See Chapter 20.

Smoke. Smoke rising in a thin vertical column indicates fair weather. Low rising or "flattened out" smoke indicates stormy weather.

Birds and Insects. Birds and insects fly lower to the ground than normal in heavy, moisture-laden air. This indicates that rain is likely. Most insect activity increases before a storm, but bee activity increases before fair weather.

Low Pressure. Slow-moving or imperceptible winds and heavy, humid air often indicate a low pressure front. This is a promise of bad weather that will probably linger for several days. You can "smell" and "hear" low pressure: The sluggish, humid air makes wilderness odors more pronounced than during high pressure. In addition, sounds are sharper and carry farther in low pressure than high pressure.

Chapter 16

Sea Survival

About 75 percent of the earth's surface is covered by water, about 70 percent being oceans and seas. You can assume, therefore, that as a soldier you will sometime cross vast expanses of water. And there is always the possibility that the plane or ship you are on will become crippled, requiring everyone to take to life rafts or lifeboats.

Precautionary Measures

As a survivor on the open sea, you will face waves and wind. You may also face extreme heat or extreme cold. To keep these environmental hazards from becoming serious problems, you must take precautionary measures as soon as possible. Use the resources available to protect yourself from the elements and from heat or extreme cold and humidity.

But protecting yourself from the elements is meeting only one of your basic needs. You must also be able to obtain water and food. Having these three basic needs will help prevent serious physical and psychological problems; however, you must know how to treat health problems that may result from your situation.

Your survival at sea depends upon—

- Your knowledge of and ability to use the survival equipment that is available.
- Your special skills and ability to apply them to cope with the hazards you face.
- Your will to live.

When you board a ship or aircraft, find out what survival equipment is on board, where it is located, and what it consists of. For instance, how many life preservers and lifeboats or rafts are on board? Where are they located? What type of survival equipment do they contain? How much food, water, and medicine do they contain?

If you are responsible for other personnel on board, make sure you know where they are and make sure they know where you are.

Down At Sea

If you are in an aircraft that goes down at sea, take the following actions once you clear the aircraft.

Whether you are in the water or in a raft—

- Get clear of and upwind of the aircraft as soon as possible, but stay in the vicinity until the aircraft sinks.
- Get clear of fuel-covered water in case the fuel becomes ignited.
- Try to find other survivors.

If you are in the water, make your way to a raft. If no rafts are available, try to find a large piece of floating debris to cling to. Relax. A person who knows how to relax in ocean water is in very little danger of drowning. The body's natural buoyancy will keep at least the top of the head above water, but some movement is needed to keep the face above water.

Floating on the back takes the least energy. Lie on your back in the water, place your arms along the sides of your body, and fin with your hands. Your head will be partially submerged but your face will be above water.

Another method for staying afloat is to float face down on the surface with the arms outstretched and the legs pointed toward the bottom. To breathe, push down on the water with your hands and raise your head above the water. Take a breath, then lower your head and return your arms to the outstretched position.

The following types of swimming strokes are recommended for a survival situation:

- Dog paddle. This is an excellent stroke when clothed or wearing a lifejacket. Although slow in speed, it requires very little energy.

- Breast stroke. This stroke should be used for swimming underwater, through oil or debris, or in rough seas. It is probably the best stroke for long range swimming in that it allows the swimmer to conserve his energy, yet maintain a reasonable speed.
- Side stroke. This is a good relief stroke in that only one arm is required to maintain momentum and buoyancy.
- Back stroke. This is an excellent relief stroke. It relieves the muscles that are used for other strokes. Use it if an underwater explosion is likely.

If you are in an area where surface oil is burning—

- Discard your shoes and buoyant life jacket.

NOTE: *If you have an uninflated* CO^2 *life jacket, keep it.*

- Cover your nose, mouth, and eyes and quickly go underwater.
- Swim underwater as far as possible before surfacing to breathe.
- To surface to breathe, force the upper part of your body above the surface. Make wide sweeping movements with your hands to splash water to disperse the flames as you surface. Try to face downwind before inhaling.
- Submerge feet first and continue as above until clear of the flames.

If the water is oil covered but free of fire, hold your head high to keep the oil out of your eyes. Attach your life preserver to your wrist and then use it as a raft.

If you should get a cramp in your legs or in your stomach, stretch and massage the cramped muscle until the cramp is gone. If cramping is severe, however, you may be unable to stretch the muscle.

If you are in a raft—

- Check the physical condition of all on board. Give first aid if necessary. Take seasickness pills if available. Vomiting, whether from seasickness or other causes, increases the danger of dehydration.
- Try to salvage all floating equipment: rations, canteens, thermos jugs, other containers, clothing, seat cushions,

parachutes, and anything else that will aid you. Secure the salvaged items in or to your raft. Make sure the items have no sharp edges that can puncture the raft.

- If there are other rafts, lash the rafts together but so they are about 25 feet apart. Be prepared to draw them closer together if you see or hear an aircraft. It is easier for an aircrew to spot rafts close together than scattered.
- Locate the emergency radio and get it into operation. Operating directions are on it. Use the emergency transceiver only when friendly aircraft are likely to be in the area.
- Have other signaling devices ready for instant use. If you are in enemy territory, avoid using a signaling device that will alert the enemy.
- Check the raft for inflation, leaks, and points of possible chafing. Make sure the main buoyancy chambers are firm (well rounded but not drum tight). Regularly check inflation. Hot air expands, so on hot days release some air. Add air when the weather cools.
- Throw out the sea anchor or improvise a drag from the raft case, bailing bucket, or roll of clothing. A sea anchor will help you stay close to your ditching site, making it easier for a searcher to find you if you have relayed your location. Wrap the sea anchor rope with cloth so it will not chafe the raft. The anchor will also help to keep the raft headed into the wind and waves.
- In stormy weather, rig the spray and windshield at once. In a 20-man raft, keep the canopy erected at all times. Keep your raft as dry as possible. Keep it properly balanced. All men should stay seated, the heaviest men in the center.

If you are in a cold climate—

- Put on an antiexposure suit. If none are available, put on extra clothing if available. Keep clothes loose and comfortable.
- Take care not to snag the raft with shoes or sharp objects. Keep the repair kit where you can readily reach it.
- Rig a windbreak, spray shield, and canopy.

- Try to keep the floor of the raft dry. Cover it with canvas or cloth for insulation.
- Huddle with others to keep warm, moving enough to keep the blood circulating. Spread an extra tarpaulin, sail, or parachute over the group.
- Give extra rations, if available, to men suffering from exposure to cold.

If you are in a hot climate—

- Rig a sunshade or canopy. Leave enough space for ventilation.
- Cover your skin where possible to protect it from sunburn. Use sunburn cream, if available, on all exposed skin. Your eyelids, the back of your ears, and the skin under your chin sunburn easily.

Calmly consider all aspects of your situation and determine what you and your companions must do to survive.

Inventory all equipment, food, and water. Waterproof items that are affected by saltwater. These include compasses, watches, sextant, matches, and lighters.

Ration water and food.

Assign duties to each person; for example, water collector, food collector, lookout, radio operator, signaler, and water bailers. Keep in mind—and remind others—that cooperation is one of the keys to survival.

Keep a log. Record the navigator's last fix, the time of ditching, the names and physical condition of personnel, ration schedule, winds, weather, direction of swells, times of sunrise and sunset, and other navigational data.

If you are down in unfriendly waters, take special security measures to avoid detection. Generally, it is best not to travel in the day time. Throw out the sea anchor and wait for nightfall before paddling or hoisting sail. Keep low in the raft; stay covered with the blue side of the camouflage cloth up. Be sure a passing ship or aircraft is friendly or neutral before you try to attract its attention. If you are detected by the enemy, destroy log book, radio, navigation equipment, maps, signaling equipment, and firearms. Jump overboard and submerge if the enemy starts strafing.

Food and Water

Obtaining food and potable water at sea may pose more problems than on land, but you can use some of the same techniques.

Food. Most fish in the open sea are edible (see page 7-18 for fish with poisonous flesh). By improvising hooks (page 7-3) and lines you can catch all the fish you need. You can use shoelaces, parachute suspension lines, or thread from clothes for a line. Small fish gather underneath the shadow of the raft. Catch them to use for bait to catch larger fish. Make a dip net to scoop up fish, crabs, and shrimp. At night some fish, especially flying fish, may land in your raft. Use the fish for food. At night shine a flashlight on the water to attract fish. Moonlight reflected on the water using a mirror will also attract fish. (Read Chapter 7 on catching and preparing fish.)

Birds are sometimes attracted to a raft for a possible perching place. If a bird lands on the raft, wait until it folds its wings before you try to grab it. Sometimes you can catch a bird with a baited hook.

Seaweed can sometimes be found floating many miles from shore. It is a rich source of minerals, but can act as a violent laxative if your stomach is not conditioned to it. Eat only small amounts at a time.

Water. If your raft contains a still, read the instructions and set it up immediately. If more than one still is available, set up as many as you need to take care of all the people in the rafts. Secure the stills to the raft with a stout line.

If desalting kits are also available, use them only for immediate water needs when stills cannot be used.

Keep a tarpaulin ready for catching rainwater. If the tarpaulin is encrusted with dried salt, wash it in sea water. There will not be enough salt left on it to harm you. At night, secure the tarpaulin like a sunshade and turn up its edges so it will collect dew. When there is rain, catch and drink as much rainwater as you can hold.

When your water supply is limited, use it efficiently.

At-Sea Medical Problems and Their Treatment

At sea you may become seasick, get saltwater sores, or face some of the same medical problems that occur on land, such as dehydration or sunburn. These problems can become critical if left untreated.

Seasickness is the nausea and vomiting caused by the motion of the raft. It can result in—

- Extreme fluid loss and exhaustion.
- Loss of will to survive.
- Others becoming seasick.
- Attracting sharks to the raft.
- Unclean conditions.

To treat seasickness—

- Wash both the patient and the raft to remove the sight and odor of vomitus.
- Keep the patient from eating food until his nausea is gone.
- Have the patient lie down and rest.
- Give the patient seasickness pills if available.

NOTE: *Some survivors have said that erecting a canopy or using the horizon as a focal point helped overcome seasickness. Others have said that swimming along side the raft for short periods helped, but extreme care must be taken if swimming.*

Saltwater sores result from a break in the skin when exposed to saltwater for an extended period of time. The sores may form scabs and pus. Do not open or drain. Flush the sores with freshwater if available and allow them to dry. Apply an antiseptic if available.

Dehydration is caused by vomiting, diarrhea, reduced water intake, sweating, and drinking alcohol, blood, urine, or seawater. Drink water at regular intervals and conserve energy.

Constipation is a common problem on a raft. Do not take a laxative, however, as this will further dehydration. Exercise as much as possible and drink an adequate amount of water if available.

Eye injuries may occur during the aircraft emergency and ditching or when you are in the water. If a foreign object becomes imbedded in an eye, bandage both eyes since movement of one eye affects the other. Do not try to remove the object.

If flame, smoke, or other contaminates get in the eyes, flush the eyes immediately with saltwater, then with freshwater if available. Apply ophthalmic ointment if available. Bandage both eyes 18 to 24 hours or longer if damage is severe. If the glare from the sky and water causes your eyes to become bloodshot and inflamed, bandage your eyes lightly. Try to prevent this problem by wearing sunglasses. Improvise sunglasses from a cloth or bandage (figure 15-3) if necessary.

Sunburn is one of the most serious problems of ocean survival. Try to prevent this problem by following the procedures on page 16-5. If you do become sunburned, treat the sunburn as any other burn

Hypothermia can occur when a person is in water that is cooler than the body temperature for long periods of time. It can also occur to a person on a raft who becomes chilled. The symptoms and treatment are described on pages 15-2 and 15-4.

Frostbite and immersion foot can also occur on a raft. Symptoms and treatment are described on pages 15-4 and 15-5.

Sea Life

Whether you are in the water or in a boat, you may see many types of sea life around you. Some may concern you more than others.

Sharks. One concern you may have is sharks. Although there are more than 325 species of sharks, only 20 or so are known to attack man. Four of the most dangerous are the great white, the mako, the tiger, and the hammerhead. Other sharks that have been known to attack man are the grey, the blue, the lemon, sand, nurse, bull, and oceanic white tip.

Some of the largest sharks—the whale shark and the basking shark, for example—are plankton feeders and of little danger. However, any shark more than 3 or 4 feet in length should be considered dangerous.

Sharks are found in all oceans and seas. Many species live and feed at considerable depths while others live and hunt near the surface.

It is the sharks that live near the surface that you are most likely to spot because of the high dorsal fins that frequently project above the water. Sharks often enter relatively shallow waters and may even enter river mouths, but are most commonly found in the open ocean or sea.

There are several kinds of sharks in the Arctic and Antarctic Oceans; however, those living in oceans and seas between 40° north and 40° south latitude are the ones most likely to be dangerous to man. Evidence shows the sharks in tropic and subtropic seas are far more aggressive than those in temperate waters.

The normal diet of most sharks is living animals. All sharks have voracious appetites. They are guided to their food by sight, scent, or sound. They will investigate floating objects and are prone to strike at injured or helpless animals, although their normal food is healthy animals. They have a highly developed sense of smell, and blood in the water excites them. They are extremely sensitive to vibrations transmitted through the water. A fish fighting against a hook and line; the struggles of an injured or dying animal; the movements of a poorly coordinated swimmer; even the rapid, excited movement of another shark approaching food will draw sharks from a greater distance than the scent of blood. They are also drawn by unusual noises such as underwater explosions.

When a shark opens its mouth and dashes forward to strike, it can bite from almost any position. It does not have to turn on its side to bite. The jaws of the great white shark and several other species are so far forward that they can bite floating objects easily without twisting to their sides.

Sharks are more apt to tear off pieces small enough to swallow easily rather than to bite off a man's leg or arm. They can and will tear great gashes in a man's torso, removing large chunks of flesh at a snap.

Most shark-inflicted injuries are half-moon incisions, often with good impressions of the smaller teeth left in the untorn flesh.

Sharks may hunt and attack singly, but most reported attacks speak of more than one present. Small sharks are likely to travel in schools and attack in numbers. A shark that finds a victim is joined at once by all sharks in the vicinity. Most sharks are cannibalistic and will attack and eat their disabled kin as readily as their prey.

Sharks are known to feed at all hours of the day and night. A review of survival accounts indicates that most shark contacts and attacks are made during daylight, a high percentage being in late afternoon. There are protective measures that you can take against sharks.

If you are in the water—

- Get near other swimmers. Several swimmers grouped together can maintain a 360° watch for sharks, can ward off sharks better than a lone swimmer, and have greater striking strength should a shark attack.
- Stay alert for sharks. Keep your clothing on, including your shoes. Evidence shows that among groups of men, those partly unclad were attacked first, and usually in the feet. In addition, clothing will protect your skin from abrasions should a shark brush against you.
- Urinate in short spurts, allowing the urine to dissipate between spurts. Pass small amounts of fecal matter at a time and throw it as far away as possible.
- If you have to vomit, vomit in your hand and throw it as far away as possible.

If you are in the water and a shark attack appears imminent—

- Make just enough commotion to keep the shark at bay. This way you conserve your strength to repulse the shark should it attack.
- Roar or yell underwater. Some divers report this will sometimes scare a shark away.
- Slap the water surface repeatedly with cupped hands. The loud sounds may scare the shark away.
- Kick and thrash at the shark. Strike the shark on the gills or eyes if you can. Hitting the shark on the nose can cause injury to your hand by glancing off the shark's nose and hitting its teeth.

If you are in a raft and you sight sharks—

- Do not fish. If you have hooked a fish before seeing the shark, let the fish go.
- Do not clean fish in the water.

- Do not throw waste overboard.
- Do not trail arms or legs in the water.
- Keep still and quiet.
- Keep hands, feet, legs, arms, and equipment inside the raft.
- Conduct all burials as soon as possible. Wait until night if sharks are numerous.

If you are in a raft and a shark attack appears imminent—

- Shoot to kill if possible.
- Fire shots at the water near the shark. The concussion may scare the shark away.
- Hit the shark with anything you have. If you use an oar, be careful not to break it. Using your hands will damage you more than the shark.

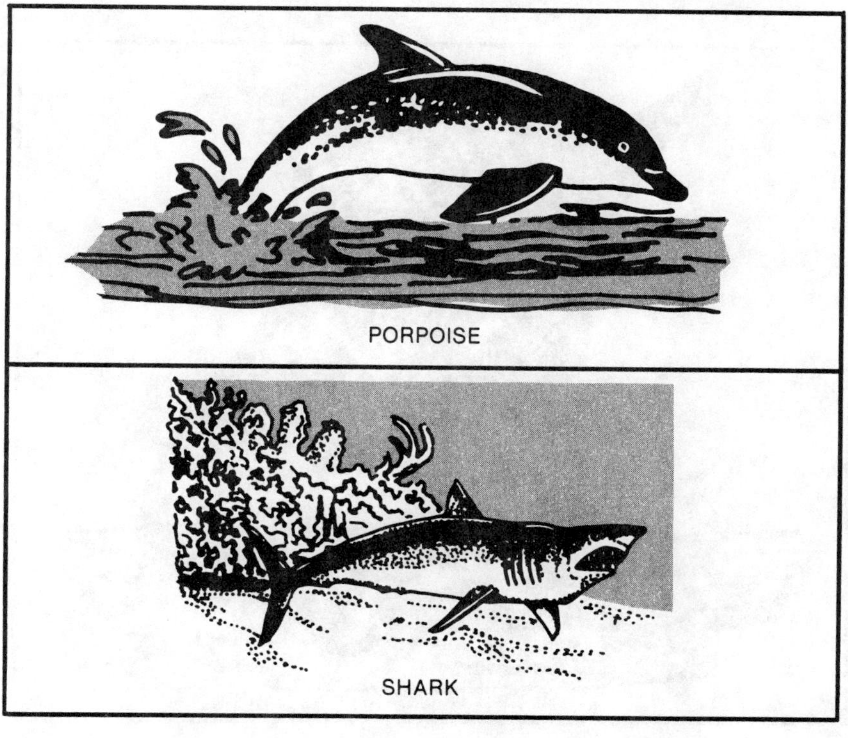

Figure 16-1. Porpoise and shark comparison.

Porpoises. Learning to tell the difference between sharks and porpoises will help to calm your fears. The porpoise may swim with its curved back or dorsal fin just above the surface. It will repeatedly dive and resurface to breathe. The shark will stay on a more continuous plane near the surface. The jumping form of the shark and porpoise also differ (figure 16-1). Observers have noted that when the porpoises appear, sharks disappear. Porpoises seem to enjoy butting rafts, but they are small enough that survivors need not be alarmed. You can drive them off by striking them with paddles.

Rays. You might also mistake the fins of the giant rays or mantas living in tropical waters for the fins of two sharks swimming side by side. If both fins periodically disappear at the same time, the fins are those of a ray (figure 16-2). In deep water all rays are harmless to swimmers. Some are dangerous, however, if stepped on in shallow water.

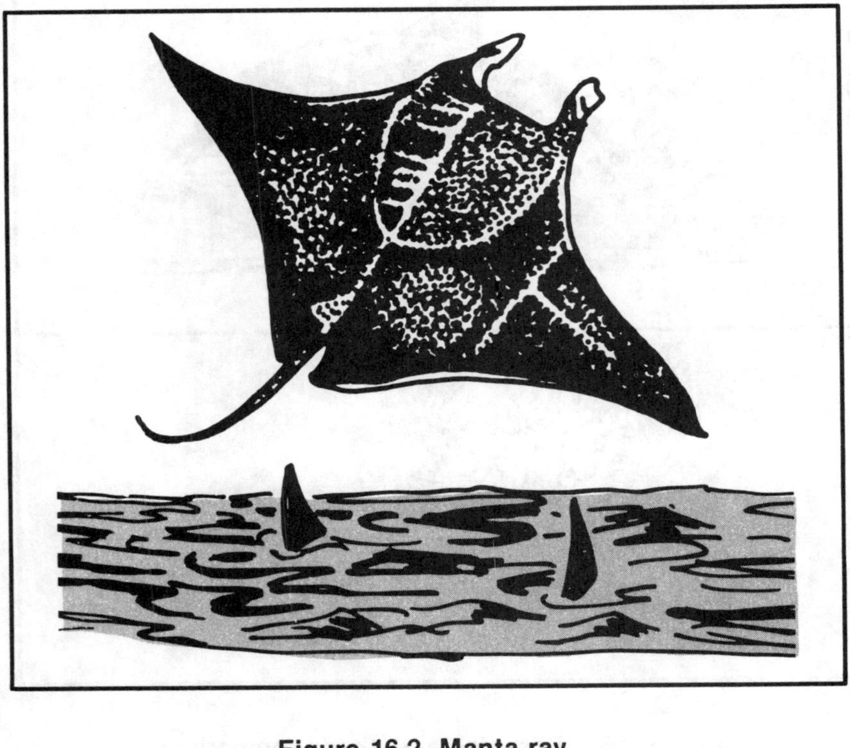

Figure 16-2. Manta ray.

Whales. The presence of whales may be disturbing because of their size. But the experience of men on rafts indicates that there is little to fear from whale attacks. Whales move leisurely when not alarmed, but their playful habit of nosing or nudging a raft may be disquieting. The deep, long drawn-out sound of their breathing and the loud sounds made by their tails slapping the water are also startling.

Raftmanship

Put out your sea anchor immediately. Do not attempt to navigate your raft unless within sight of shore or in unfriendly waters. Remember that the majority of successful rescues are made within 7 days of ditching. You can't go very far on a raft in 7 days.

Lookouts. Keep at least one lookout posted at all times to watch for signs of land, passing vessels or aircraft, wreckage, seaweed, schools of fish, birds, and signs that the raft may be chafing or leaking. Rotate the lookouts at least every 2 hours.

Traveling. Your raft will move regardless of what you do. The course it will take depends on the wind, the ocean current, and your use of oars or paddles, tiller, sea anchor, and sails.

When ocean currents are moving toward your destination, but the winds are unfavorable, put out a sea anchor. Huddle in the raft to offer as little wind resistance as possible. In the open ocean, currents seldom move more than 6 to 8 miles a day.

A raft lacks a keel, so you can't sail it into the wind. However, you can sail it downwind. Multiplace rafts (except 20-man) can be successfully sailed 10^0 from the wind direction. Don't try to sail your raft unless you know that land is near.

When the wind is blowing directly toward your destination, fully inflate the raft, sit high, take in the sea anchor, rig a sail, and use an oar as a rudder.

In a multiplace raft (except a 20-man), rig a square sail in the bow, using oars with their extensions as mast and crossbar.

If no regular sail is available, use the waterproof tarpaulin or one or two thicknesses of parachute cloth.

If the raft has no regular mast socket and step, erect the mast by tying it securely to the front cross seat; provide braces. Pad the bottom of the mast to prevent it from chafing or punching a hole through the floor. Improvise a mast step with a shoe by wedging the toe under the seat and using the heel as the mast step.

Do not secure the corner of the lower edge of the sail. Hold the lines attached to the corners in your hands so that a sudden storm or gust of wind will not rip the sail, break the mast, or capsize the raft.

Take care to prevent your raft from turning over. In rough weather, keep the sea anchor out from the bow. All passengers should sit low in the raft with their weight distributed to hold the weather side down. Do not sit on the sides or stand up. Never make sudden movements without warning the other men. Do not tie a fishline to yourself or to the raft; a large fish may capsize the raft.

In rough seas, tie stern of first raft to bow of second and rig a sea anchor to stern of second raft. Use approximately a 25-foot line between rafts; adjust the length of the line to suit the sea. Keep the sea anchor line long. Adjust its length so that when the raft is at the crest of a wave, the sea anchor will stay in a trough. In very rough weather, keep a spare sea anchor rigged for instant use in case the one in use breaks loose.

When the sea anchor is not in use, tie it to the raft and stow it so that it will hold immediately if the raft capsizes.

To right multiplace rafts (except a 20-man), use the righting handles on the bottom of the raft. Work from the bottom side of the raft. Try to maneuver so you and the bottom side are downwind. Grasp the righting handle and lift the far side of the raft up and over (figure 16-3). Both bottom and top of a 20-man raft are identical and, therefore, the raft requires no righting.

If several men are in the water, one man should hold down the far side of the multiplace raft (4- to 7-man) while the rest climb in singly from the other side. Grasp the seat to haul yourself in, or use a boarding ladder if available. Without help, the best place to board the raft is over the end. If the wind is blowing, board the raft with the wind at your back. The 20-man raft is provided with an inflated boarding station.

To board the 1-man raft, climb in from the narrow end; slide up as nearly horizontal as possible.

Figure 16-3. Righting a raft correctly.

Signs of Land

The lookout should watch carefully for signs of land.

A fixed cumulus cloud in a clear sky or in a sky where all other clouds are moving often hovers over or slightly downwind from an island.

In the tropics, a greenish tint in the sky is often caused by the reflection of sunlight from shallow lagoons or shelves of coral reefs.

In the arctic, ice fields or snow-covered land are often indicated by light-colored reflections on clouds. These reflections are quite different from the darkish gray ones caused by open water.

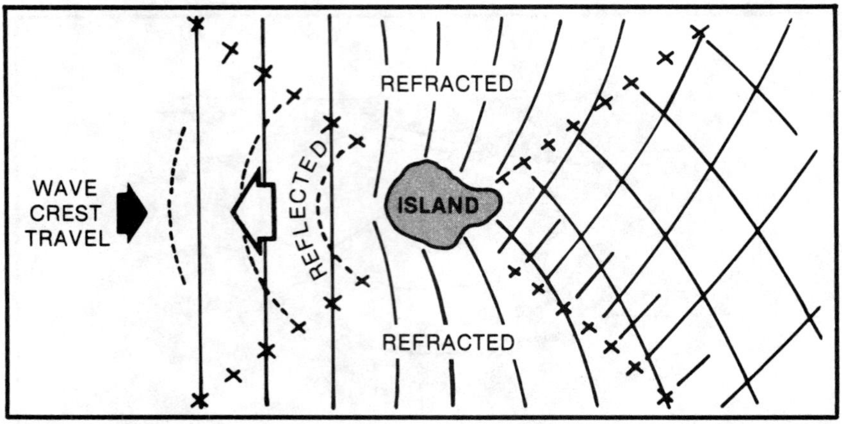

Figure 16-4. Wave patterns about an island.

Deep water is dark green or dark blue. Lighter color indicates shallow water, which may mean land is near.

In fog, mist, or rain, or at night you may detect land by odors and sounds. The musty odor of mangrove swamps and mud flats and the smell of burning wood carry a long way. The roar of surf is heard long before the surf is seen. Continued cries of sea birds from one direction indicate their roosting place on nearby land.

Usually more birds are found near land than over the open sea. The direction from which flocks fly at dawn and to which they fly at dusk may indicate the direction of land. During the day, birds are searching for food and the direction of flight has no significance.

In the tropics, mirages may be seen, especially during the middle of the day. Be careful not to mistake a mirage for nearby land. A mirage disappears or its appearance and elevation change when it is viewed from slightly different heights.

You may be able to detect land by the pattern of the waves, which are refracted as they approach land (figure 16-4). By traveling with the waves and parallel to the slightly turbulent area marked "X" on the illustration, you should reach land.

To swim ashore—

- Wear your shoes and at least one thickness of clothing. Use the side or breast stroke to conserve strength.
- If surf is moderate, ride in on the back of a small wave by swimming forward with it. Shallow dive to end your ride just before the wave breaks.

- In high surf, swim shoreward in the trough between waves. When the seaward wave approaches, face it and submerge. After it passes, work shoreward in the next trough.
- If you are caught in the undertow of a large wave, push off the bottom or swim to the surface and proceed shoreward as above.
- If you must land on a rocky shore, look for a place where the waves rush up onto the rocks. Avoid places where the waves explode with a high white spray. Swim slowly in making your approach—you will need your strength to hold on to the rocks.
- After selecting your landing point, advance behind a large wave into the breakers. Face shoreward and take a sitting position with your feet in front, 2 or 3 feet lower than your head, so that your feet will absorb shocks when you land or strike submerged boulders or reefs.
- If you don't reach shore behind the wave you have picked, swim with hands only. As the next wave approaches, take a sitting position with feet forward. Repeat procedure until you land.
- Water is quieter in the lee of a heavy growth of seaweed. Take advantage of such growth. Don't swim through the seaweed; crawl over the top by grasping the vegetation with overhand movements.
- Cross a rocky reef just as you would land on a rocky shore. Keep your feet close together and your knees slightly bent in a relaxed sitting posture to cushion blows against coral.

To raft ashore—

- You can use the 1-man raft without danger in most cases. However, going ashore in a strong surf is dangerous. Take your time. Select your landing point carefully. Try not to land when the sun is low and straight in front of you. Try to land on the lee side of an island or of a point of land. Keep your eyes open for gaps in the surf line, and head for them. Avoid coral reefs and rocky cliffs. Coral reefs do not occur near the mouths of freshwater streams. Avoid rip currents or strong tidal currents which may carry you far out to sea. Either signal shore for help or sail around and look for a sloping beach where the surf is gentle.

- If you have to go through surf to reach shore, take down the raft mast. Keep your clothes and shoes on to avoid severe cuts. Adjust and inflate your life vest. Trail the sea anchor over the stern with as much line as you have. Use the oars or paddles and constantly adjust the sea anchor to keep a strain on the anchor line. This will keep your raft pointed toward shore and prevent the sea from throwing the stern around and capsizing you. Use the oars or paddles to help ride in on the seaward side of a large wave.
- Surf may be irregular and velocity may vary, so modify your procedure as conditions demand. A good method of getting through surf is to have half the men sit on one side of the raft, half on the other, facing each other. When a heavy sea bears down, half should row (pull) toward the sea until the crest passes; then the other half should row (pull) toward the shore until the next heavy sea comes along.
- Against strong wind and heavy surf, the raft must have all possible speed to pass rapidly through the oncoming crest in order to avoid being turned broadside or thrown end over end. If possible, avoid meeting a large wave at the moment it breaks.
- In medium surf with no wind or offshore wind, keep raft from passing over a wave so rapidly that it drops suddenly after topping the crest. If the raft turns over in the surf, try to grab hold.
- As the raft nears the beach, ride in on the crest of a large wave. Paddle or row hard and ride in onto the beach as far as you can. Don't jump out of the raft until it has grounded, then quickly get out and beach it.
- If you have a choice, don't land at night. If you have reason to believe that the shore is inhabited, lay away from the beach, signal, and wait for the inhabitants to come out and bring you in.
- If you encounter sea ice, land only on large, stable floes. Avoid icebergs, which may capsize, and small floes or those obviously disintegrating. Use oars and hands to keep raft from rubbing on ice edge. Take raft out of the water and store well back from ice edge. Keep raft inflated and ready for use. Any floe may break up.

Chapter 17

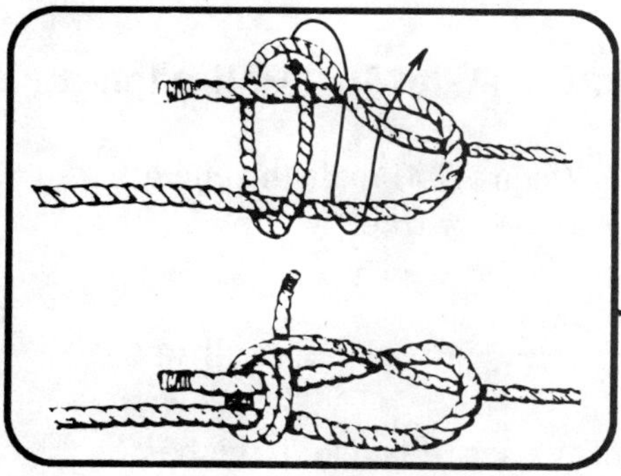

Knots

In a survival situation, you need to know how to tie different knots and what knot to use to meet a particular need. For instance, making snares, field expedient weapons, and shelters requires the use of knots.

Square Knot

Use a square knot (figure17-1) to tie the ends of two ropes of equal diameter together. A square knot tightens under strain, but is easy to untie by grasping the ends of the two bights and pulling the knot apart.

NOTE: *A square knot made using wet rope or ropes of different diameters will not hold.*

Prusik Knot

Use a prusik knot (figure17-2) for tying a sling rope to a climbing rope. The knot will hold the sling in place when tension is put on it, but will slide up or down the climbing rope when you release the tension. You can also use this knot when weaving a fishnet.

Clove Hitch

Use the clove hitch (figure17-3) to fasten a rope to a pole, post , or similar object. You can make the knot at any point on the rope; however, to make the knot hold, you must either keep tension on it or run an extra loop around the anchor object and under the center of the clove hitch.

Round Turn With Two Half Hitches

Use a round turn with two half hitches (figure17-4) to tie the end of a rope around an object such as a pipe, post, or tree.

Bowline Knot

The bowline knot (figure17-5) has many uses. It is one of the best knots for forming a single loop that will not become smaller when tension is placed on it.

Double Sheet Bend

Use a double sheet bend (figure17-6) to join two ropes of unequal diameter, two ropes that are wet, two tubular nylon cords, or two straps.

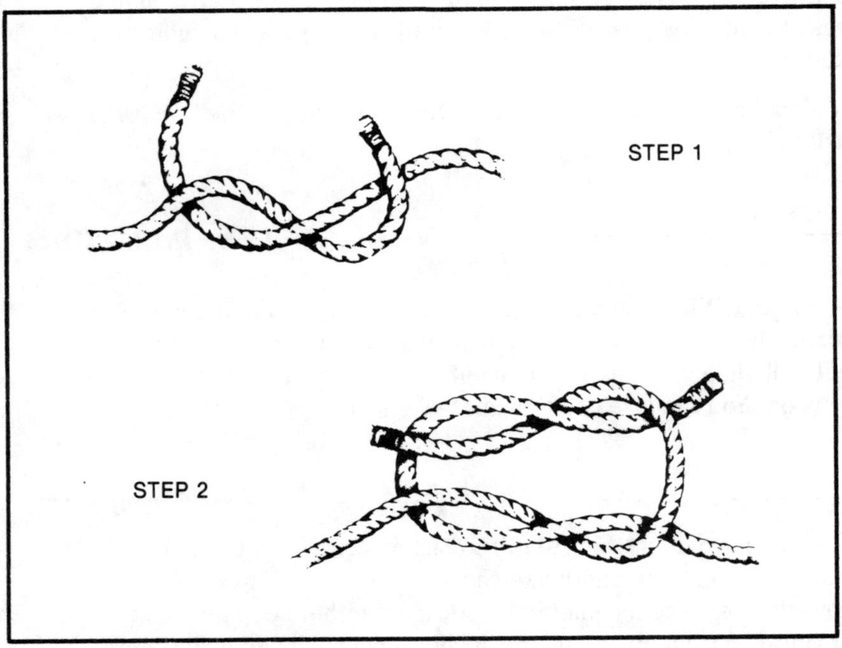

Figure 17-1. Square knot.

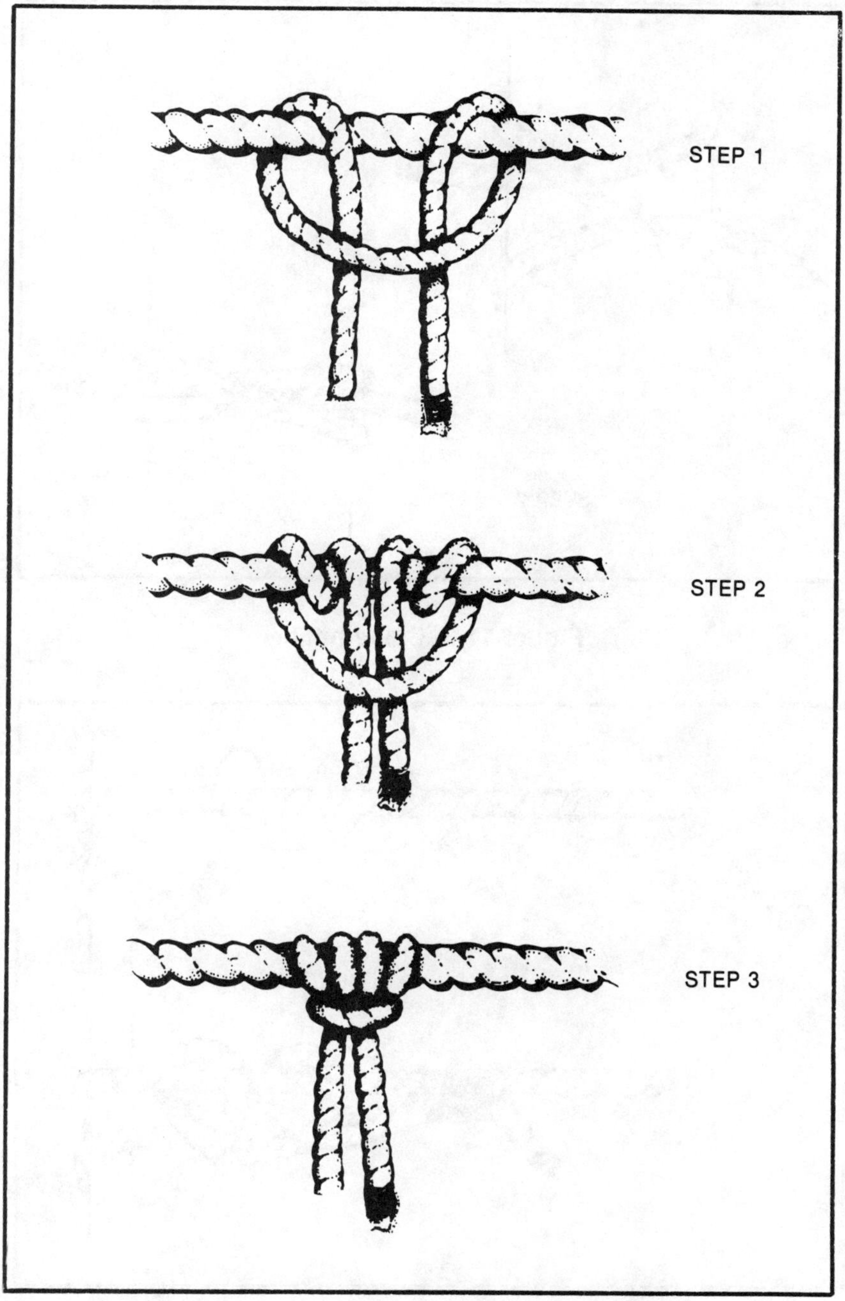

Figure 17-2. Prusik knot.

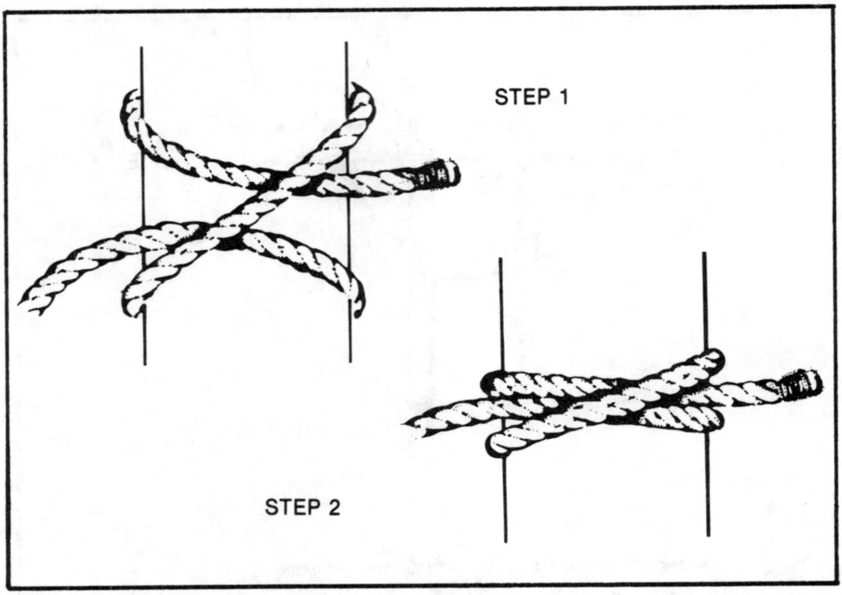

Figure 17-3. Clove hitch.

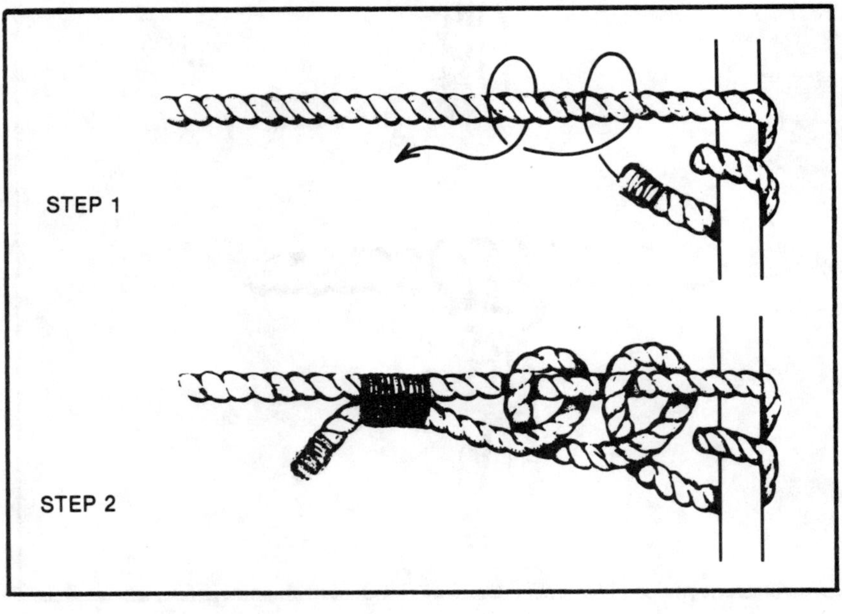

Figure 17-4. Round turn with two half hitches.

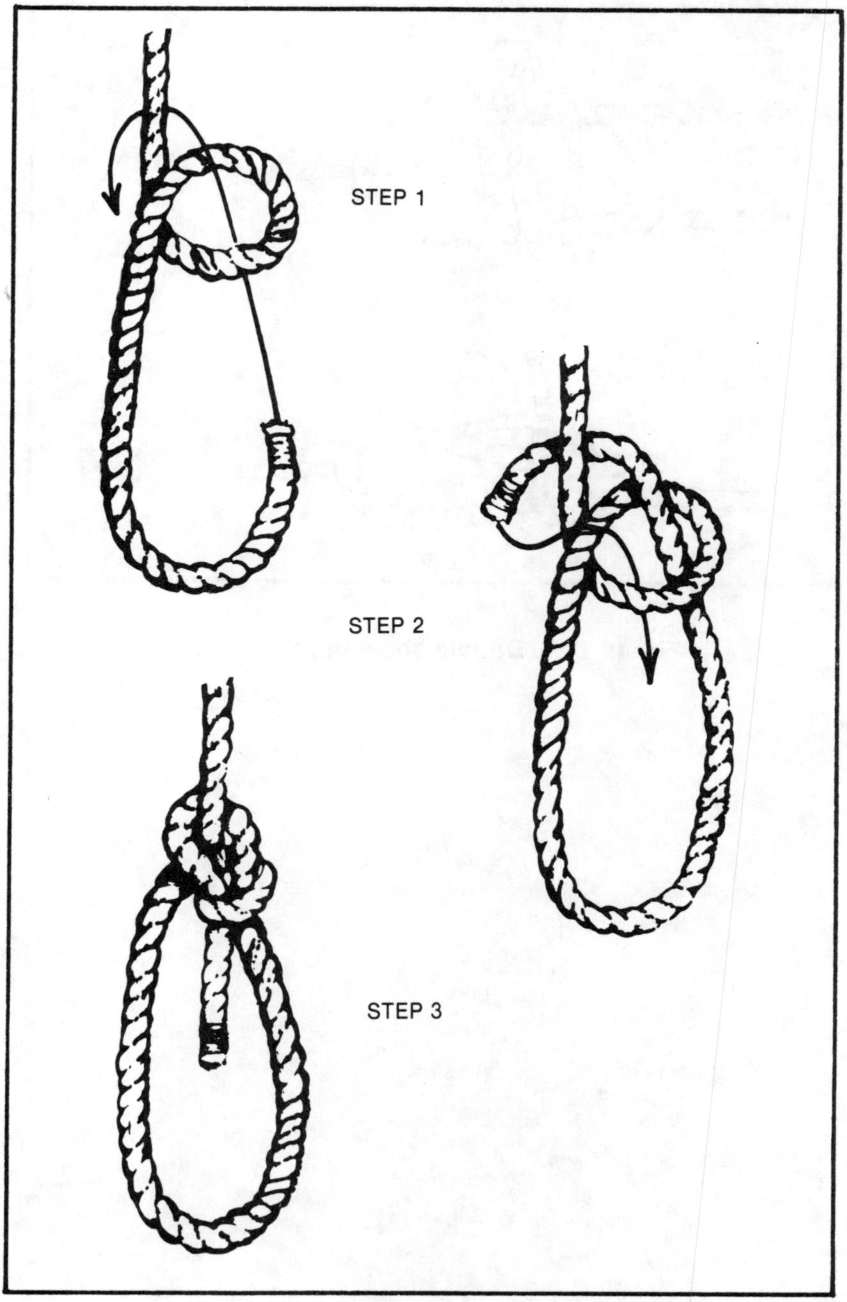

Figure 17-5. Bowline knot.

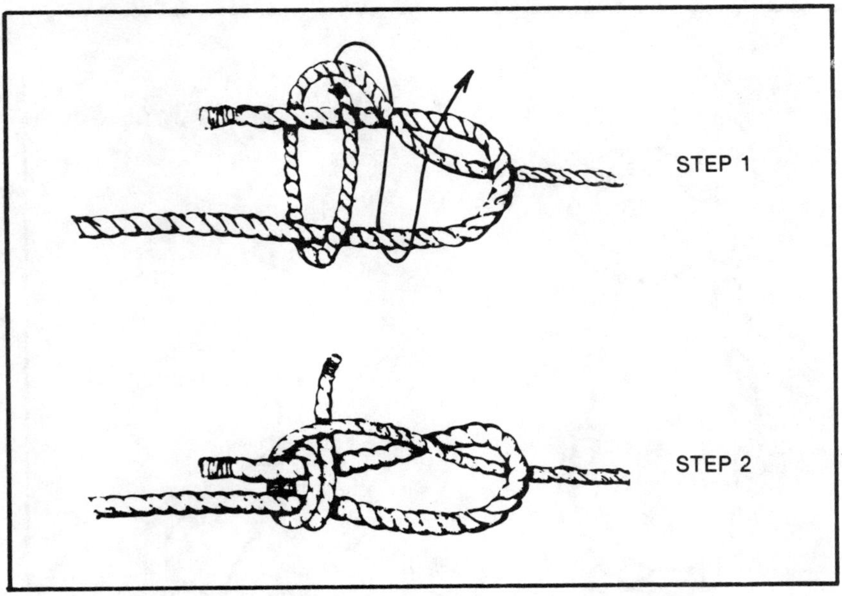

Figure 17-6. Double sheet bend knot.

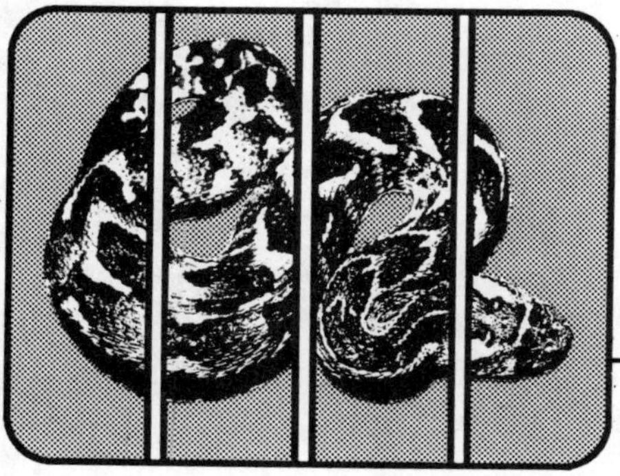

Appendix A

Poisonous Snakes

If you fear snakes, it is probably because you are unfamiliar with them or you have wrong information about them. There is no need for you to fear snakes if you know—

- *their habits,*
- *how to identify the dangerous kinds,*
- *precautions to take to prevent snakebite, and*
- *what actions to take in case of snakebite (Chapter 2).*

For a man wearing shoes and trousers and living in a camp, the danger of being bitten by a poisonous snake is small compared to the hazards of malaria, cholera, dysentery, or other diseases.

Nearly all snakes avoid man if possible. Reportedly, however, a few—the king cobra of southeast Asia, the bushmaster of South America, the tropical rattlesnake of South America, and the mamba of Africa—sometimes aggressively attack man, but even these snakes do so only occasionally. Most snakes get out of the way and are seldom seen.

Ways to Avoid Snakebite

Most bites occur when a person accidentally steps on a snake, so keep your eyes open. Wear shoes and trousers when you are in areas where poisonous snakes are found.

Take the following precautions to avoid snakebite:

- Avoid areas where snakes are abundant if possible.
- Use a snake stick or walking stick if in a snake-infested area.

- Probe before entering an area.
- Don't chase snakes or tease them.
- Don't put your hands in a place you cannot see, such as in a hole, in bushes, or under a river bank.
- Don't handle a snake unless its head is cut off or unless you are sure it has been dead for some time.
- Watch for snakes.
- Learn as much as possible about snakes in your area (types, habits, size, shape, color, and so forth).

Snake Groups

Snakes dangerous to man generally fall into two groups: proteroglypha and solenoglypha. These two groups are best described in terms of their fangs and their venom.

Fangs. The proteroglypha have in front of the upper jaw and preceding the ordinary teeth permanently erect fangs. These fangs are referred to as fixed fangs.

The solenoglypha have erectile fangs; that is, the fangs are capable of being raised to an erect position. These fangs are referred to as folded fangs.

Venom. The fixed-fang snakes (proteroglypha) usually have neurotoxic venoms. These venoms affect the nervous system, making the victim unable to breathe.

The folded-fang snakes (solenoglypha) usually have hemotoxic venoms. These venoms affect the circulatory system, destroying blood cells, damaging skin tissues, and causing internal hemorrhaging.

It is important to note, however, that most poisonous snakes have both neurotoxic and hemotoxic venom. But usually one type of venom in the snake is dominant and the other is weak.

Poisonous versus nonpoisonous snakes. No single characteristic distinguishes a poisonous snake from a harmless one except the presence of poison fangs and glands. And only in dead specimens can you determine the presence of these fangs and glands without danger.

Family and Common Name	Group		Venom	
	Folded Fangs	Fixed Fangs	Neurotoxic	Hemotoxic
Viperidae or true vipers				
European long nosed viper	●			●
Gaboon viper	●		●	●
Puff adder	●			●
Rhinoceros viper	●		●	●
Russell's viper	●			●
Sand viper	●			●
Crotalids or pit vipers				
American copperhead	●			●
Bushmaster	●			●
Cottonmouth	●			●
Eastern diamondback rattlesnake	●			●
Fer-de-lance	●			●
Green tree pit viper	●			●
Habu pit viper	●			●
Jumping pit viper	●			●
Malayan pit viper	●			●
Tropical rattlesnake	●		●	●
Wagler's pit viper	●			●
Western diamondback rattlesnake	●			●
Mojave rattlesnake	●		●	●
Elapids				
Common cobra		●	●	
Common krait		●	●	
Coral snake		●	●	
Death adder		●	●	
Egyptian cobra		●	●	
King cobra		●	●	
Tiger snake		●	●	
Australian copperhead		●	●	
Mamba		●	●	
Hydrophids				
Sea snakes		●	●	

Figure A-1. Families of poisonous snakes.

There are many different poisonous snakes throughout the world, and it's unlikely you will see many except in a zoo. Only a few poisonous snakes are described in this manual.

Viperidae (True Vipers). The viperidae or true vipers generally have thick bodies and heads that are much wider than their necks. However, there are many different sizes, markings, and colorations.

Crotalids (Pit Vipers). The crotalids or pit vipers may be either slender or thick-bodied. Their heads are usually much wider than their necks. These snakes take their name from the deep pit located between the eye and the nostril. They are commonly brown with dark blotches, though some kinds are green.

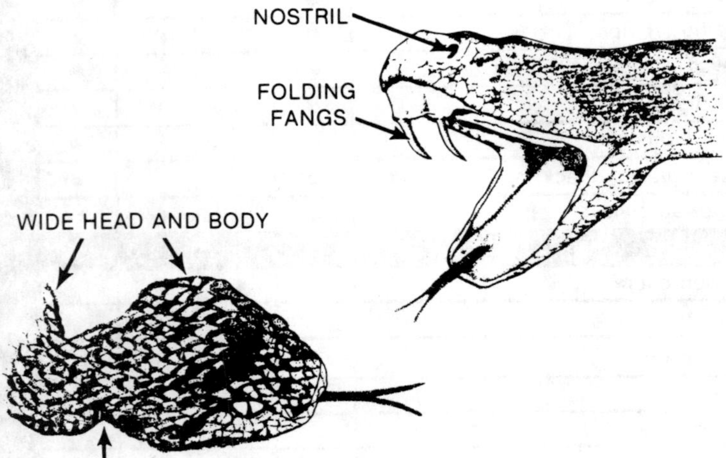

Figure A-2. Positive identification of vipers.

India has about 12 species of these snakes. They can be found in trees or on the ground in all types of terrain. The tree snakes are slender; the ground snakes are heavy-bodied. All are dangerous.

China has a pit viper similar to the moccasin found in North America. It is found in the rocky areas of the remote mountains of South China. It reaches a length of 4-1/2 feet, but is not vicious unless irritated. A small pit viper, about 1-1/2 feet long, is found on the plains of eastern China. It is too small to be dangerous to a man wearing shoes.

There are about 27 species of rattlesnakes in the United States and Mexico. They vary in color and may or may not have spots or blotches. Some are small while others, such as the diamond-backs, may grow to 8 feet long.

There are five kinds of rattlesnakes in Central and South America, but only the tropical rattlesnake is widely distributed. The rattle on the tip of the tail is sufficient identification for a rattlesnake.

Most will try to escape without a fight when approached, but there is always a chance one will strike at a passerby. They do not always give a warning; they may strike first and rattle afterwards or not at all.

Elapids. Only by examining a dead snake can you positively determine if it is a cobra or a near relative (figureA-4). On cobras, kraits, and coral snakes, the third scale on the upper lip touches both the nostril scale and the eye. The krait also has a row of enlarged scales down the ridged back.

Figure A-3. Positive identification of pit vipers.

The cobras of Africa and the near East can be found in almost any habitat. One kind may live in or near water, another may be found in trees. Some are reported to be aggressive and savage. The distance a cobra can strike in a forward direction is equal to the distance its head is raised above the ground. Some cobras, however, can spit venom a distance of 10 to 12 feet. This venom is harmless unless it gets into your eyes; then it may cause blindness if not washed out immediately.

Hydrophids. There are many species of sea snakes. They vary greatly in the color and shape but all have flat tails. Their scales distinguish them from eels, which have no scales.

Sea snakes occur in saltwater along the coasts throughout the Pacific. There are also sea snakes on the east coast of Africa and in the Persian Gulf. None are found in the Atlantic Ocean.

There is no need to fear sea snakes. They have not been known to attack a man swimming. Fishermen occasionally get bit by a sea snake caught in a net. The bite is dangerous.

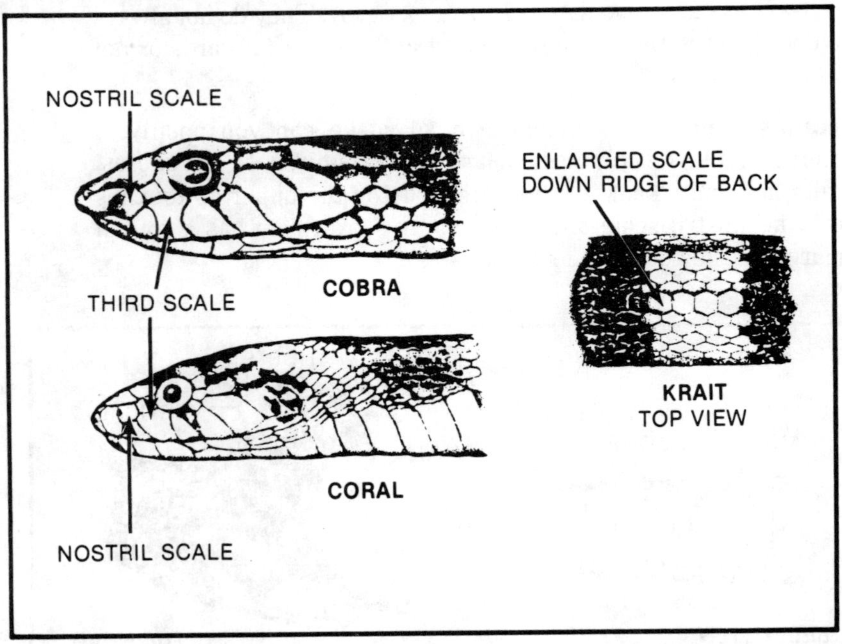

Figure A-4. Positive identification of cobras, kraits, and coral snakes

Courtesy of Andy Koukoulis, Florida Cypress Gardens

European long nosed viper *(Vipera ammodytes)*

Description: Coloration is gray with a zigzag pattern running the length of the back. A few snakes are solid black. This is a small viper, usually about 1-1/2 to 2 feet. *The venom is hemotoxic.* The bite is very painful, causing considerable tissue damage. Although the bite is serious, it is rarely fatal. Bites should be medically treated.

Characteristics: Has a short temper and readily strikes.

Habitat: Many terrains, including swamps, open fields, hillsides, and farming regions. It can be found above 9,000 feet elevation in mountainous regions.

Distribution: Throughout northern Europe to northern Korea.

Figure A-5. European long nosed viper.

Courtesy of Andy Koukoulis, Florida Cypress Gardens

Gaboon viper *(Bitis gabonica)*

Description: Coloration is striking with geometric patterns of blacks, tans, and blues. Colors are so complex that these vipers are difficult to see among the forest flora. This is the largest and heaviest of all the vipers. Adults average 4 to 5 feet, but have been known to reach over 6 feet. These vipers usually have horns on the tip of the nose; however, the horns may be absent on some specimens. Their fangs are enormous—over 2 inches long in large vipers. *Their venom contains both hemotoxin and neurotoxin.* Without prompt medical treatment, bites are usually lethal.

Characteristics: When approached, will coil and strike quickly.

Habitat: Heavy forests.

Distribution: Tropical rain forests of Sierra Leone, Sudan, Angola, and Natal.

Figure A-6. Gaboon viper.

Courtesy of Andy Koukoulis, Florida Cypress Gardens

Puff adder *(Bitis arietans)*

Description: Coloration is variable, usually light or dark brown with white or yellow chevrons. Adults average 3 to 4 feet, but occasionally reach 5 feet. The body appears heavy, often giving the impression of sluggishness. Adult puff adders have long hollow fangs. *Their venom is very hemotoxic.* Without proper medical treatment, death can occur.

Characteristics: When approached, this snake is quick to react, coiling and hissing loudly. It strikes with lightning speed.

Habitat: Open forests or grasslands near streams.

Distribution: Most of Africa, except the arid hot region of the desert and the tropical rain forest.

Figure A-7. Puff adder.

Courtesy of Andy Koukoulis, Florida Cypress Gardens

Rhinoceros viper (also called river jack) *(Bitis nasicornis)*

Description: Coloration is variable with many hues of pinks, blues, greens, and purples. The pattern is quite striking. These are fairly large, heavy bodied vipers with long horns on the nose. Adults average 2 to 3 feet in length, but have been known to reach 4 feet. *The venom contains two toxins: hemotoxin and neurotoxin.*

Characteristics: When approached, it is quick to react, coiling, and hissing loudly, and striking with lightning speed.

Habitat: Tropical rain forest.

Distribution: Liberia, Uganda, and Zaire.

Figure A-8. Rhinoceros viper.

Courtesy of Andy Koukoulis, Florida Cypress Gardens

Russell's viper *(Vipera russellii)*

Description: Coloration is tan or brownish yellow with oval spots ringed with black. Average size is 3-1/2 to 5 feet. This is a very common and dangerous snake over much of its range. It is responsible for many snakebites and deaths. *Its venom is a very strong hemotoxin.*

Characteristics: Bold disposition; hisses loudly and is ready to strike with great speed.

Habitat: Varies from farm lands to prairies to swamps. Prefers open, sunny spots, but can be found almost anywhere except in thick jungle.

Distribution: West Pakistan, all of India, Burma, Thailand, and southwest China.

Figure A-9. Russell's viper.

Courtesy of Andy Koukoulis, Florida Cypress Gardens

Sand viper *(Cerastes vipera)*

Description: Coloration is yellow or pinkish to match the sand. These snakes are ordinarily 2 feet or less in length. *The venom is hemotoxic.*

Characteristics: The body is capable of flattening out and burrowing into the sand. These snakes have a gliding gait, similar to the sidewinder rattlesnake, enabling them to move swiftly across the desert sands.

Habitat: Dry, sandy desert areas.

Distribution: Northern Africa from Algeria to Egypt.

Figure A-10. Sand viper.

Courtesy of Andy Koukoulis, Florida Cypress Gardens

American copperhead *(Agkistrodon contortrix)*

Description: Usually pale brown with darker crossbands narrowing at the midline of the back. The markings may be few and inconspicuous on large snakes. The head is usually coppery-red in color. The belly is light colored and somewhat mottled. *The venom is hemotoxic,* but is weak and not particularly dangerous to adults.

Characteristics: Is timid and will usually stay hidden; will try to escape when discovered. If cornered, may vibrate tail and produce a distinct buzzing sound.

Habitat: Prefers high, dry ground. May be found in the fields or woods.

Distribution: Southern and southeastern United States.

Figure A-11. American copperhead.

Courtesy of Andy Koukoulis, Florida Cypress Gardens

Bushmaster *(Lachesis muta)*

Description: Coloration is tan or dark brown with some pink hues and with black blotches along the back. The scales are extremely keeled. This is the largest of all pit vipers, averaging 6-1/2 to 7 feet, with some recorded up to 12 feet. It has long hollow fangs and *hemotoxic venom.* Snakebite of humans seldom occurs with this species, but a bite is serious.

Characteristics: May remain motionless until touched; may attempt to escape when cornered and may strike viciously. It has no rattle but its tail vibrating against dry leaves produces a rattling sound.

Habitat: Mostly in forests at low altitudes; however, may be found in cool areas. Prefers moist to dry ground and often hides in animal burrows.

Distribution: Southern Nicaragua, Costa Rica, Ecuador, Peru, Bolivia, Brazil, and Paraguay.

Figure A-12. Bushmaster.

Courtesy of Andy Koukoulis, Florida Cypress Gardens

Cottonmouth *(Agkistrodon piscivorus)*

Description: Coloration is variable. The young are brilliantly colored with bands of copper and light and dark browns. In adults, the bands may be faint to total black. Adults average 3 to 4 feet, but occasionally reach 6 feet. It has *hemotoxic venom.* Deaths are rare but do occur.

Characteristics: A good swimmer. Often basks on logs or leaning trees in swamps, bayous, or along sluggish streams. Usually will retreat when disturbed but may hold its ground.

Habitat: Swampy areas, lakes, rivers, or streams.

Distribution: Throughout southern United States, southern Illinois, Missouri, southeast Kansas, central Texas, Virginia, the Carolinas, and all of Florida.

Figure A-13. Cottonmouth.

Courtesy of Andy Koukoulis, Florida Cypress Gardens

Eastern diamondback rattlesnake *(Crotalus adamanteus)*

Description: Coloration is olive green with dark diamond-shaped markings with white or yellow edgings. This is the largest venomous snake in the United States, averaging 3-1/2 to 5 feet in length; maximum length recorded is over 7 feet. It has large venom glands and long hollow fangs. *The venom is hemotoxic.*

Characteristics: Bold. Will readily defend itself by coiling, inflating its body with air making a low pronounced hiss; rattle will usually vibrate.

Habitat: Prefers open, sandy places or rocky ledges, but may be found in practically any type of terrain.

Distribution: North Carolina southward to Florida and west to Louisiana.

Figure A-14. Eastern diamondback rattlesnake.

Courtesy of Andy Koukoulis, Florida Cypress Gardens

Fer-de-lance *(Brothrops lanceolatus)*

There are several closely related species in this group. All are very dangerous to man.

Description: Coloration is variable, but usually brown or olive with faint hourglass markings along the back. Adult snakes average 3-1/2 to 5 feet, but 7-foot specimens have been recorded. *The venom is strongly hemotoxic.* This species is responsible for many snakebite fatalities.

Characteristics: May coil its body before striking; however, can strike from any position.

Habitat: The large kinds are ground snakes, and are often found in cane fields or around dwellings. Some of the small kind live in trees, especially at the base of palm trees.

Distribution: Throughout Central and South America, except Chile.

Figure A-15. Fer-de-lance.

Courtesy of Andy Koukoulis, Florida Cypress Gardens

Green tree pit vipers *(Trimeresurus sp.)*

Description: These are small snakes, brilliant green in color. Bites are common but seldom fatal. *The venom is hemotoxic.* Bites from these snakes cause considerable pain and local tissue damage. Bites usually occur on the upper extremities, neck, head or shoulders.

Characteristics: Lives primarily in shrubs or trees.

Habitat: Commonly found on tea plantations in thick brush.

Distribution: Cambodia, Burma, Malaysia, Indonesia, Vietnam, India, Sunda Islands, and S.E. China.

Figure A-16. Green tree pit viper.

Courtesy of Andy Koukoulis, Florida Cypress Gardens

Habu pit viper *(Trimeresurus flavoviridis)*

Description: Coloration is light brown with long dark brown to greenish blotches edged with yellow. Adults average 4 to 5 feet in length; maximum length is 7 feet. *Venom is hemotoxic of low toxicity;* deaths have been recorded.

NOTE: The high incidence of snakebite by this species makes it important. It often enters human dwellings in search of rodents.

Characteristics: Irritable disposition, striking readily.

Habitat: Found in many environments.

Distribution: Okinawa, Amami.

Figure A-17. Habu pit viper.

Courtesy of Andy Koukoulis, Florida Cypress Gardens

Jumping pit viper *(Bothrops nummifera)*

Description: This is a robust snake, short and husky. Average length is 2½ to 3 feet. The coloration is from a rich brown to black. *The venom is hemotoxic, though death is rare. The bites are painful and local tissue destruction is extensive.* The appearance of this snake is similar to the dreaded bushmaster, having exceedingly rough scales.

Characteristics: Strikes with such force that its body will at times leave the ground.

Habitat: Found in many different environments.

Distribution: Southern Mexico through Central America.

Figure A-18. Jumping pit viper.

Courtesy of Andy Koukoulis, Florida Cypress Gardens

Malayan pit viper *(Callaselasma rhodostoma)*

Description: A small snake, 2 to 3 feet long. Coloration is a reddish brown back with dark brown crossbands, narrow at the backbone, pinkish brown on the sides. *The venom is hemotoxic and very painful and damaging to tissue.* Hundreds of bites a year are attributed to this species, but the death rate is low.

Characteristics: Has a calm disposition, but will bite if stepped on.

Habitat: It's common on rubber plantations.

Distribution: Thailand, Laos, Cambodia, Java, Sumatra, Malaysia, and Vietnam.

Figure A-19. Malayan pit viper.

Courtesy of Andy Koukoulis, Florida Cypress Gardens

Tropical rattlesnake *(Crotalus durissus)*

Description: Coloration is dark brown with darker diamond-shaped markings along the back. Long dark stripes run along the neck. These snakes average 4 to 5 feet in length, but some may reach 6 feet. *The venom contains considerable neurotoxic elements, causing systemic problems, but very little local tissue damage.* Symptoms include dizziness, blindness, difficult breathing, and paralyzed neck muscles.

Characteristics: Seldom uses its rattle for warning. When approached, this snake will usually coil and elevate its head high above the coil; it may use the rattle to sound a few "clicks."

Habitat: Found in dry, hilly country.

Distribution: Dry areas of southern Mexico and all of Central and South America except Chile.

Figure A-20. Tropical rattlesnake.

Courtesy of Andy Koukoulis, Florida Cypress Gardens

Wagler's pit viper (also called temple viper) *(Trimeresurus wagleri)*

Description: Coloration is green with scattered black-edged scales. It has a rather stout body; adults average 2 to 3 feet in length. *It has hemotoxic venom, which is fairly toxic and considered dangerous.*

Characteristics: Rather placid disposition; will bite if molested.

Habitat: Rain forest environment. Often found around human settlements.

NOTE: In some villages, natives consider finding this species close to home is a good omen.

Distribution: Thailand, Malaysia, Indonesia, Borneo, and Philippines.

Figure A-21. Wagler's pit viper.

Pit Vipers

Western diamondback rattlesnake *(Crotalus atrox)*

Description: Coloration is light brown with darker brown diamond-shaped markings; tail is cream to white, heavily barred with black rings. This is the second largest venomous rattlesnake in the United States, averaging 3 to 5 feet; maximum length recorded is 7 feet. *The venom is hemotoxic, causing considerable pain and tissue damage.*

Characteristics: Bold. Will readily defend itself.

Habitat: Prefers open, sandy places or rocky ledges, but may be found in practically any type of terrain.

Distribution: Texas, Louisiana, Arkansas, southeast California, northern Mexico.

Figure A-22. Western diamondback rattlesnake

Pit Vipers

Mojave rattlesnake *(Crotalus scutulatus)*

Description: Coloration is greenish or olive, similar to the western diamondback rattlesnake. Adults average 2-1/2 to 3 feet; maximum length is 4 feet. It has *both hemotoxic and neurotoxic venom.* This is an important species due to its highly toxic venom, which has a more marked effect on respiration than any other North American rattlesnake. Bites have little local tissue damage or reaction and may not look serious until respiratory difficulty begins.

Characteristics: Does not always give a warning. May strike first and rattle afterwards or not at all.

Habitat: Arid desert and prairie desert environments.

Distribution: West Texas, Mojave Desert in California into the Mexican highlands.

Figure A-23. Mojave rattlesnake.

Courtesy of Andy Koukoulis, Florida Cypress Gardens

Common cobra *(Naja naja)*

Description: Coloration of adults is brown or black with bicolor scales. There is often a "spectacle" mark on the back of the hood. Average length is 4 to 5 feet; maximum is 6-1/2 feet. *Venom is neurotoxic.*

Characteristics: Ready to defend itself at all times. Raises the forward part of the body one-half to two-thirds off the ground, spreading a wide hood. It strikes forward and downward.

Habitat: Lives in many environments, often entering human settlements in search of rodents.

Distribution: All of India, Sri Lanka, Pakistan, Burma, Thailand, Cambodia, Vietnam, Laos, Malaysia, Sumatra, Java, Celebes.

Figure A-24. Common cobra.

Courtesy of Andy Koukoulis, Florida Cypress Gardens

Common krait *(Bungarus caeruleus)*

Description: Coloration is grayish to black with narrow white crossbands. Belly is white. Average length is 3 to 6 feet. *The venom is a potent neurotoxin.* Death rate from snakebite is very high.

Characteristics: Nocturnal. Is not aggressive. When alarmed, the body will jerk in different directions, raising the tail off the ground.

Habitat: Open country. Often found near inhabited areas. Many have been found in poor settlement buildings.

Distribution: India and West Pakistan.

Figure A-25. Common krait.

Courtesy of Andy Koukoulis, Florida Cypress Gardens

Coral snake *(Micrurus fulvius)*

Description: Coloration is vivid blacks, yellows, and reds. The head is small, the body cylindrical. Average length is 1-1/2 to 3 feet. Coral snakes have two tiny fixed fangs. They do not strike, but will often bite when touched or stepped on. Chewing is not required to release the venom. *The venom is a very potent neurotoxin,* Symptoms may not appear for several hours after the bite. Take the victim to a medical facility, observe, and start medical treatment. Antivenin is available for this species.

Characteristics: Secretive habits. Not aggressive; will not bite unless picked up or stepped on. Active in early morning and evening hours.

Habitat: Moist wooded areas. Also found in suburban areas looking for food.

Distribution: Southern United States (coastal North Carolina to west Texas).

Figure A-26. Coral snake.

Courtesy of Andy Koukoulis, Florida Cypress Gardens

Death adder *(Acanthophis antarcticus)*
Although the name implies that this snake is an adder, it is not a viperidae; it belongs to the cobra family (elapidae).

Description: Coloration is light brown, reddish with darker crossbands. Its tail is yellow and short. This species is small, averaging 18 inches, seldom reaching 3 feet. It has a short, thick body, a head much wider than its neck, and a short, thin tail. *Its venom is extremely neurotoxic.* There is about a 50% mortality rate if bites are not medically treated.

Characteristics: Short temper. Bites quickly with little provocation.

Habitat: Sandy localities.

Distribution: New Guinea, Ceram, Tanimbar, Islands of the Bismark Archipelago, Australia, Schouten.

Figure A-27. Death adder.

Courtesy of Andy Koukoulis, Florida Cypress Gardens

Egyptian cobra *(Naja haje)*

Description: Coloration is variable. Adults may be black, brown, or yellowish; some are light brown with darker brown wide crossbands. Average length is 5 to 6 feet; maximum is 8 feet. This species is rather heavy bodied and has a very wide long hood, which it displays when alarmed. ***Its venom is a very potent neurotoxin.***

Characteristics: When molested, is ready to defend itself and may be aggressive.

Habitat: Cultivated lands, rocky hillsides, old ruins, and rural villages in arid regions.

Distribution: Most of northern Africa and southwestern Arabian Peninsula.

Figure A-28. Egyptian cobra.

Courtesy of Andy Koukoulis, Florida Cypress Gardens

King cobra *(Ophiophagus hannah)*

Description: Coloration is olive, light brown; some may have faint darker cross bands. This is the largest of all venomous snakes. The average length is 7 to 9 feet, but a few have been recorded to 18 feet. Large cobras can stand 3 to 4 feet off the ground with the hood extended, hissing loudly, almost like a growl. *The venom is a powerful neurotoxin.* Without immediate medical aid, death is certain.

Characteristics: May attack deliberately, especially if guarding eggs.

Habitat: Primarily jungle, but can be found near inhabited areas.

Distribution: Thailand, India, southeast China, Philippines, Malaya, Burma, Pakistan, Cambodia, Laos, Vietnam, Sumatra, Java, Borneo, and Celebes.

Figure A-29. King cobra.

Courtesy of Andy Koukoulis, Florida Cypress Gardens

Tiger snake *(Notechis scutatus)*

Description: Coloration is variable from greenish gray to yellowish to total black. Most have narrow dark bands across the back. Adult snakes average 3-1/2 to 5 feet; a few have been recorded to 7 feet. *This is one of the most deadly snakes in the world. The venom is neurotoxic.* Death usually occurs swiftly if victim is not treated.

Characteristics: Aggressive when molested. Spreads its neck when angry. Quick to bite, lunging with a flashing stroke so vigorous that body sometimes moves forward. Active at night. Humans fall victim to the bite by stepping on the snake while walking along roads.

Habitat: Dry country.

Distribution: Tasmania, southern Australia to Queensland, Bass Straits Island.

Figure A-30. Tiger snake.

Courtesy of Waikiki Aquarium, Honolulu, Hawaii

Yellow-bellied sea snake *(Pelamis platurus)*

Description: Coloration is black to dark brown on top; bright yellow on the sides and belly. Averages 25 to 30 inches in length. Tail is flattened like an oar. *Venom is neurotoxic.* Fatalities have been recorded.

Distribution: Many of the Pacific Islands to Hawaii and to the coast of Costa Rica, and Panama.

Figure A-33. Yellow-bellied sea snake.

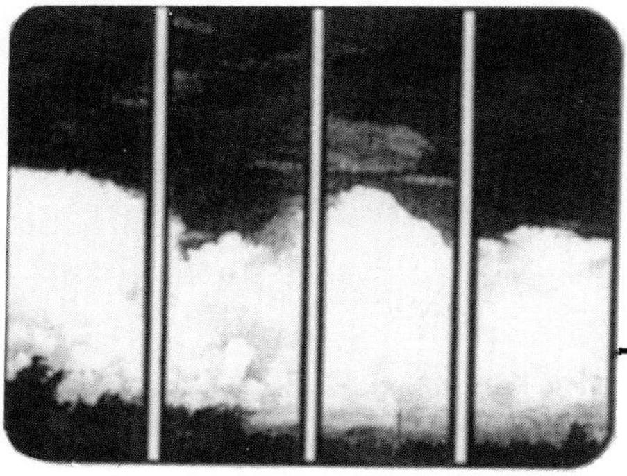

Appendix B

Clouds—Foretellers of Weather

About 200 years ago an Englishman classified clouds according to what they looked like to a person seeing them from the ground. He grouped them into three classes and gave them Latin names: cirrus, cumulus, *and* stratus. *These three names, alone and combined with other Latin words, are still used to identify different cloud formations.*

By being familiar with the different cloud formations and what weather they portend, you can take appropriate action for your protection.

Cirrus Clouds

Cirrus clouds (figure B-1) are the very high clouds that look like thin streaks or curls. They are usually 4 miles or more above the earth and are usually a sign of fair weather. In cold climates, however, cirrus clouds that begin to multiply and are accompanied by increasing winds blowing steadily from a northerly direction indicate an oncoming blizzard.

Cumulus Clouds

Cumulus clouds (figure B-2) are fluffy, white, heaped-up clouds. These clouds, which are much lower than cirrus clouds, are often fair weather clouds. They are apt to appear around midday on a sunny day, looking like large cotton balls with flat bottoms. As the day advances, they may become bigger and push higher into the atmosphere, piling up to appear like a mountain of clouds. These can turn into storm clouds.

Stratus Clouds

Stratus clouds (figure B-3) are very low gray clouds, often making an even gray layer over the whole sky. These clouds generally mean rain.

Nimbus Clouds

A nimbus (figure B-4) is a rain cloud of uniform greyness that extends over the entire sky.

Cumulonimbus Clouds

Cumulonimbus (figure B-5) is the cloud formation resulting from a cumulus cloud building up, extending to great heights, and forming in the shape of an anvil. You can expect a thunderstorm if this cloud is moving in your direction.

Cirrostratus Clouds

Cirrostratus is a fairly uniform layer of high stratus clouds (figure B-6) that are darker than cirrus. Cirrocumulus is a small, white, round cloud at a high altitude. Cirrostratus and cirrocumulus clouds indicate good weather.

Scuds

A loose, vapory cloud (scud) driven before the wind is a sign of continuing bad weather.

Courtesy of National Oceanic and Atmospheric Administration

Figure B-1. Cirrus clouds.

Courtesy of National Oceanic and Atmospheric Administration

Figure B-2. Cumulus clouds.

Courtesy of National Oceanic and Atmospheric Administration

Figure B-3. Stratus clouds.

Courtesy of National Oceanic and Atmospheric Administration

Figure B-4. Nimbus clouds.

Courtesy of National Oceanic and Atmospheric Administration

Figure B-5. Cumulonimbus clouds.

Courtesy of National Oceanic and Atmospheric Administration

Figure B-6. Cirrostratus clouds.

Edible Plants

If you are in a situation where you must find edible plants for food, you must know where to look for them and how to identify them.

Habitats and Climatic Zones

Knowing what edible plants grow in your area and climatic zone will help you in your search for plant food. In this appendix the terms temperate, tropical, and arctic are used to describe the climatic zones (in terms of temperature range). Keep in mind, however, that the temperate and tropical zones have mountainous and desert areas. At high elevations in the mountains of temperate and tropical zones, the temperature range is comparable to that in the arctic. And in desert areas in these zones, the temperature range is comparable to that of the tropics.

In this appendix, the following terms are used to describe the habitats and climatic zones in which plants grow:

- Tundra—a treeless area in arctic regions (or at high elevations in mountains in temperate or tropical regions) often with frost only a few feet below the soil.
- Aquatic area—a water-covered or marshy area.
- Forest—a dense growth of trees and underbrush.
- Grassland—a treeless plain or prairie dominated by grasses.
- Disturbed area—an area where the natural vegetation has been removed or altered.
- Cultivated area—an area used for growing crops.
- Savanna—an area with drought-resistant grasses and a few scattered trees.

- Desert scrub—a desert area that is dominated by shrubs.
- Human settlement—an area around a village, a house, or houses or structures in which people live.

If you know where to look for edible plants and can identify them you should be able to "live off the land" in a survival situation.

Identifying Plants

As you look at the pictures and read the descriptions of plants in this appendix, keep in mind their leaf margins, shapes, and arrangements and their root structures. This will help you to identify plants wherever you are.

Leaf Characteristics

The leaf margins may be toothed, entire (toothless), or lobed (figure C-1). Toothed leaves may be fine toothed, coarse toothed, single toothed, or double toothed. Entire leaves have no divisions, lobes, or teeth. Some leaves may have wavy edges.

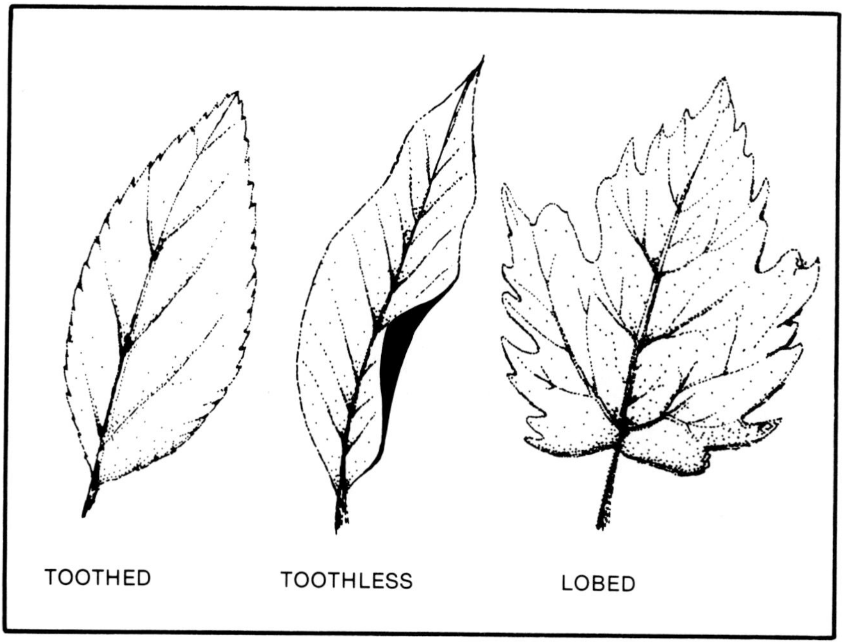

Figure C-1. Leaf margins.

Leaves may be lance-shaped, elliptic, egg-shaped, oblong, wedge-shaped, triangular, long pointed, or top-shaped (figure C-2).

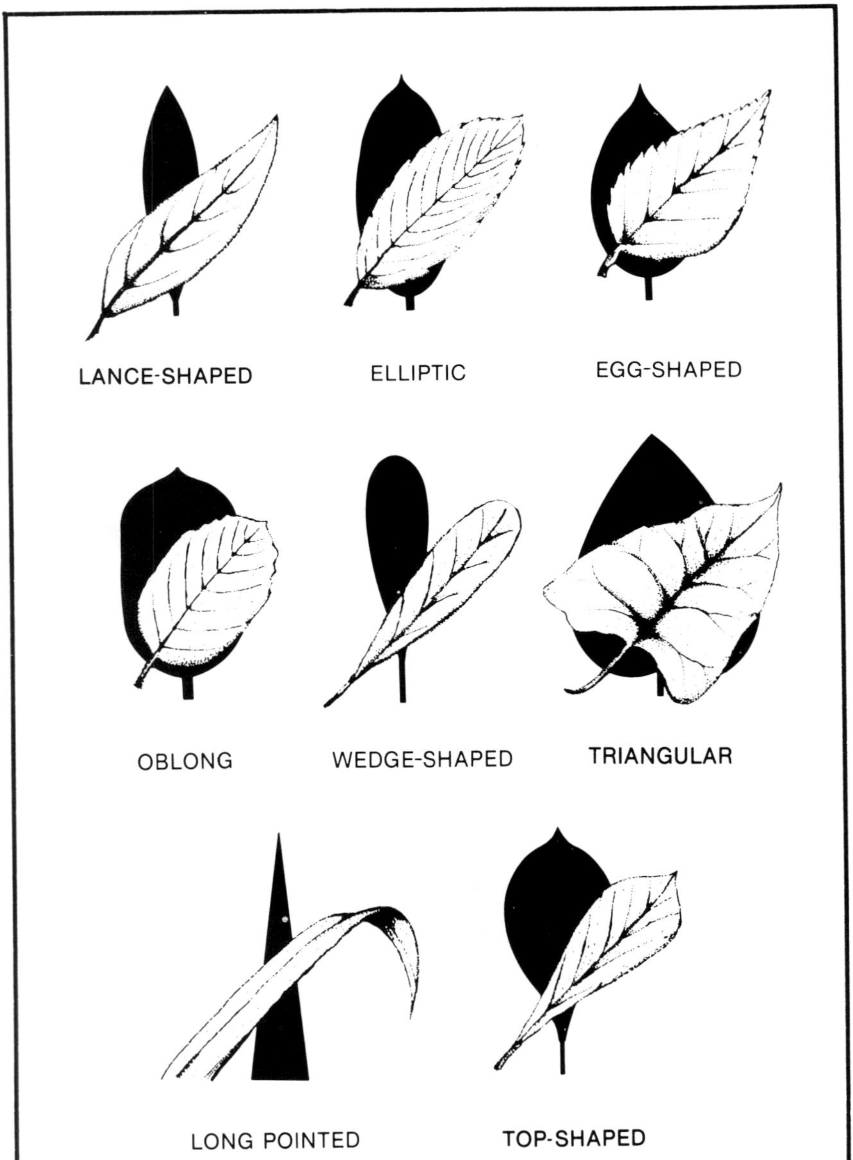

Figure C-2. Leaf shapes.

The basic types of leaf arrangements are opposite, alternate, compound, simple, and basal rosette (figure C-3).

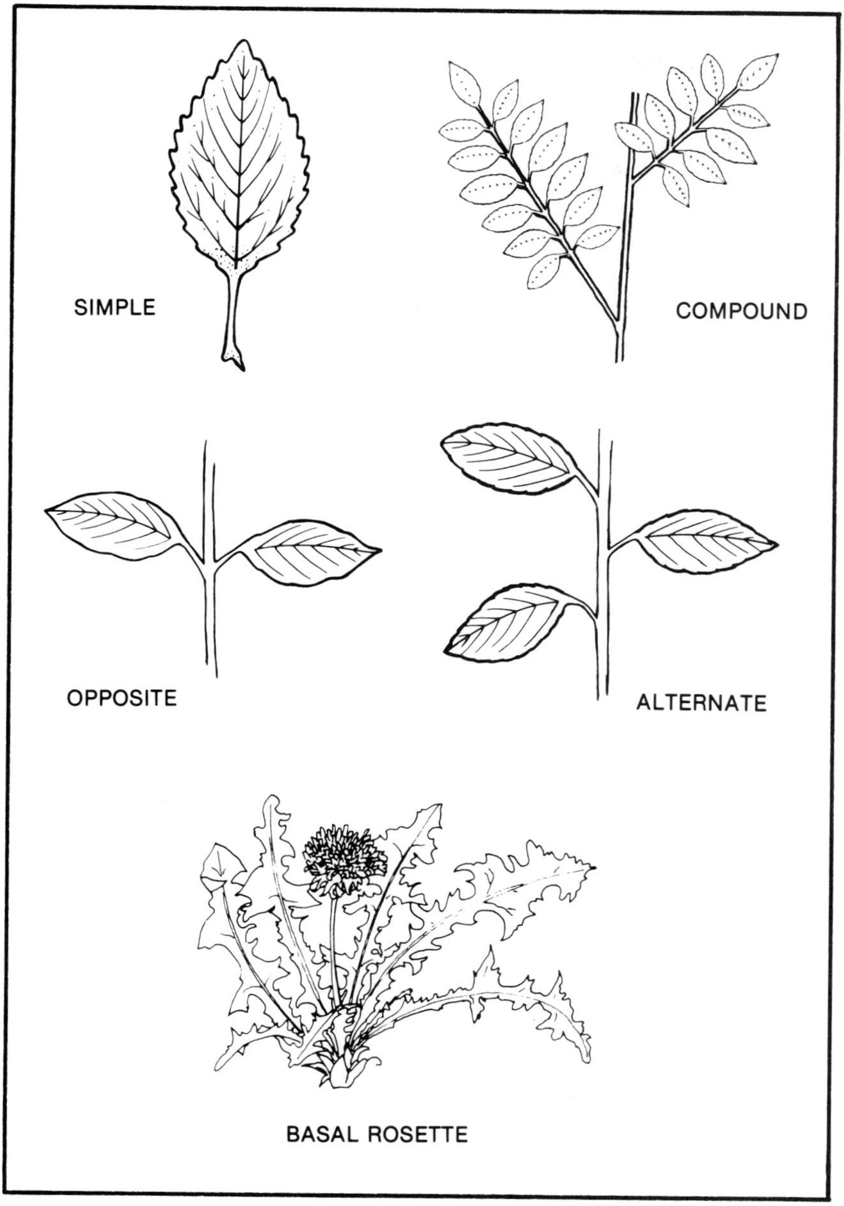

Figure C-3. Leaf arrangements.

Root Structures. The four basic types of root structures are bulb, taproot, tuber, and rhizome (figure C-4).

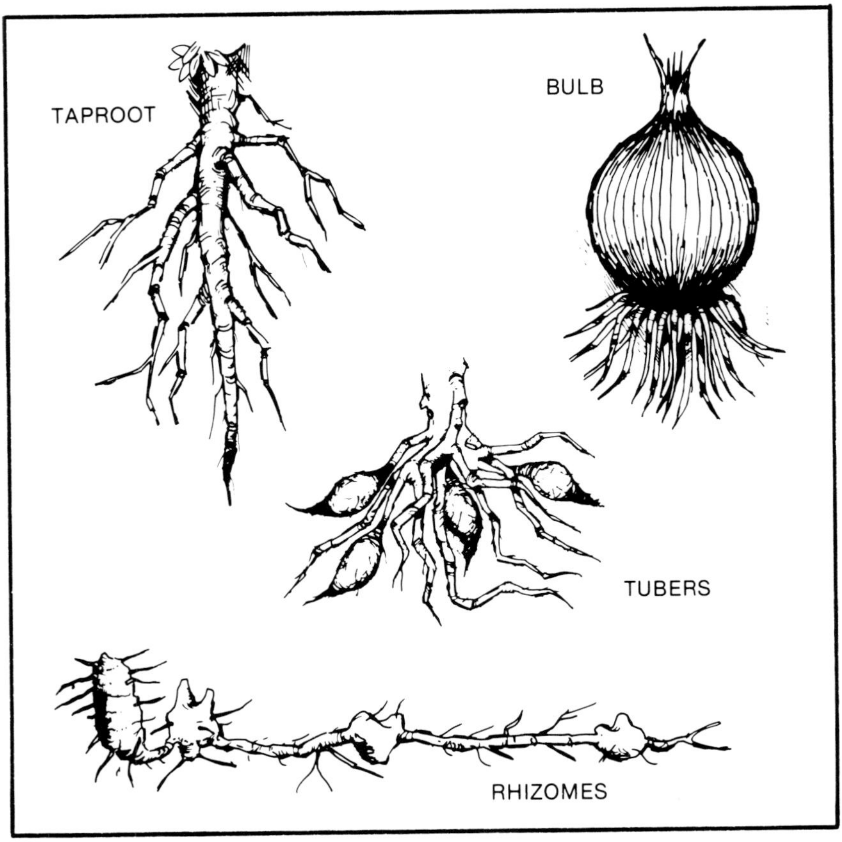

Figure C-4. Root structures.

Selecting and Preparing Plants

In addition to the information in Chapter 5 on edibility and preparation, you should keep the following in mind when you are looking for plant food:

- There are no known poisonous grasses, so you should consider wild grains as an important emergency food source. Grains are edible raw, but roasting or boiling improves the food quality.

CAUTION: Many grasses are infected by the ergct fungus, which is extremely toxic. Ergot forms a dark black hornlike growth on the grain. This growth is about l inch long and persists through winter. Do not *eat any plant that has this fungus on it.*

- Some cacti are *not* edible. If the cactus grows close to the ground (not more than 5 inches tall) and does not have padlike segments, avoid it.
- The young leaves of all ferns uncoil from a fiddleheadlike structure, and it is at this stage that ferns are best for eating.

CAUTION: Limit the amount of fern leaves you eat as some species contain a material that may destroy vitamins in the body.

- Eating large quantities of some moss (lichen) may cause sickness.
- Water lilies, cattails, and water fern—although edible raw—should be washed in water suitable for drinking since the water they grow in may be contaminated.
- The roots of some root crops (taro, arrowroot, malanga, yam) should not be eaten raw as most contain mildly poisonous compounds. Boiling or baking destroys these compounds. It also destroys harmful bacteria that may be present on the root surface.
- Many trees have edible leaves, flowers, and roots, as well as edible fruit. Some trees may be a source of water. Most are an excellent resource for building materials.
- If you are unsure of the edibility of a plant, apply the Universal Edibility Test.

Photographs and Descriptions

There are many wild plants that have edible parts. A few of these plants are shown/described in figures C-5 through C-78. Plants with both photographs and descriptions are placed in alphabetical order so that you can quickly flip to them. Plants with descriptions only are listed in figure C-79.

As you look at the figures, note the habitats and distributions of these plants to find out if they grow in your area. Learn to spot and identify them immediately. Keep in mind, however, that in another area the same plants, certainly different species of the same plants, may vary in size and structure.

Acacia *(Acacia farnesiana)*

Description: This is a spreading, usually short tree with spines and alternate compound leaves. The individual leaflets are small. The flowers are ball-shaped, bright yellow, and very fragrant. The bark is a whitish grey color. The fruits are dark brown and podlike.

Habitat and distribution: Acacia grows in open, sunny areas. It is found throughout all tropical regions.

NOTE: There are about 500 kinds of acacia. These plants are especially prevalent in Africa, southern Asia, and Australia, but many kinds occur in the warmer and drier parts of America.

Edible parts: The young leaves, flowers, and pods are edible raw or cooked.

Figure C-5. Acacia.

Agave *(Agave species)*

Description: These plants have large clusters of thick, fleshy leaves borne close to the ground and surrounding a central stalk. The plants flower only once, then die. They produce a massive flower stalk.

Habitat and distribution: Agaves prefer dry, open areas. They are found throughout Central America, the Caribbean, and parts of the western deserts of the United States and Mexico.

Edible parts: The flowers and flower buds are edible. Boil before eating.

CAUTION: The juice of some species causes dermatitis.

Other uses: Cut the huge flower stalk and collect the juice for drinking. Some species have very fibrous leaves. Pound the leaves and remove the fibers to use for ropes and weaving. Most species have thick, sharp needles at the tips of the leaves. These can be used for sewing or making hooks. The sap of some species contains a chemical that makes it suitable for use as a soap.

Figure C-6. Agave

Arrowroot *(Maranta arundinacea)*

Description: This plant grows up to 5 foot tall. Its leaves are 1 foot long and 4 inches wide. The leaves fold at night.

Habitat and distribution: This plant is a native of South America but is now grown on a wide scale in the humid tropics. Look for it in open sunny areas.

Edible parts: The rootstock is a rich source of high quality starch. Boil the rootstock and eat it as a vegetable.

Figure C-7. Arrowroot.

Bamboo *(various genera including Bambusa, Dendrocalamus, Phyllostachys)*

Description: Bamboos are woody grasses that grow up to 50 feet tall. The leaves are grasslike and the stems are the familiar bamboo used in furniture and fishing poles.

Habitat and distribution: Look for bamboo in warm, moist regions in open or jungle country, in lowland, or on mountains. Bamboos are native to the Far East, both temperate and tropical zones, but have been widely planted around the world.

Edible parts: The young shoots of almost all species are edible raw or cooked. Raw shoots have a slightly bitter taste that is removed by boiling. To prepare, remove the tough protective sheath, which is coated with tawny or red hairs. The seed grain of the flowering bamboo is also edible. Boil the seeds like rice or pulverize them, mix with water, and make into cakes.

Other uses: Use the mature bamboo to build structures or to make containers, ladles, spoons, and various other cooking utensils.

Bamboo can also be used to make tools and weapons (see Chapter 4). You can make a strong bow by splitting the bamboo and putting several pieces together.

Figure C-8. Bamboo.

Bananas and plantains *(Musa species)*

Description: These are treelike plants with several large leaves at the top. The flowers are borne in dense hanging clusters.

Habitat and distribution: Look for bananas in open fields or margins of forests where they are grown as a crop. They grow in the humid tropics.

Edible parts: The fruits are edible raw or cooked. They may be boiled or baked. The flowers can be boiled and eaten like a vegetable. The rootstalks and leaf sheaths of many species can be cooked and eaten. The center or "heart" of the plant is edible year round, cooked or raw.

Other uses: Layers of the lower third of the plant can be used to cover coals to roast food. The stump of the plant can be used to obtain water (page 5-7).

Figure C-9. Bananas.

Baobab *(Adansonia digitata)*

Description: Baobab trees may grow as high as 60 feet and may have a trunk 30 feet in diameter. They have short stubby branches and a grey, thick bark. The leaves are compound and the segments are arranged like the palm of a hand. The flowers, which are white and several inches across, hang from the higher branches. The fruits are shaped like a football, up to 1½ feet long, and covered with short dense hair.

Habitat and distribution: These trees grow in savannas. They are found in Africa, in parts of Australia, and on the island of Madagascar.

Edible parts: The young leaves can be used as a soup vegetable. The tender root of the young baobab tree is edible. The pulp and seeds of the fruit are also edible. Use one handful of pulp to about one cup of water for a refreshing drink. To obtain flour, roast the seeds, then grind them.

Other uses: Drinking a mixture of pulp and water will help cure diarrhea.

Often the hollow trunks are good sources of freshwater.

The bark can be cut into strips and pounded to obtain a strong fiber for rope.

Figure C-10. Baobab.

Batoko plum *(Flacourtia inermis)*

Description: This shrub or small tree has dark green, alternate, simple leaves. Its fruits are bright red and contain six or more seeds.

Habitat and description: This plant is a native of the Philippines but is widely cultivated for the fruits. It can be found in clearings and at the edges of the tropical rain forests of Africa and Asia.

Edible parts: Eat the fruits raw or cooked.

Figure C-11. Batoko plum.

Bearberry, sometimes called Kinnikinnick *(Arctostaphylos uvaursi)*

Description: This is a common evergreen shrub with reddish, scaly bark and thick, leathery leaves 1½ inches long and ½ inch wide. It has white flowers and bright red fruits.

Habitat and distribution: This plant is found in arctic and subarctic regions most often in sandy or rocky soil.

Edible parts: The berries are edible raw or cooked. The young leaves can be used to make a refreshing tea.

Figure C-12. Bearberry.

Blackberries, raspberries, and dewberries *(Rubus)*

Description: These plants have prickly stems (canes) that grow upward, arching back toward the ground. They have alternate, usually compound leaves. The fruits may be red, black, yellow, or orange in color.

Habitat and distribution: These plants grow in open sunny areas at the margin of woods, lakes, streams, and roads throughout temperate regions.

Edible parts: The fruits and peeled young shoots are edible.

Other uses: Use the leaves to make tea. To treat diarrhea, drink a tea made by brewing the dried root bark of the blackberry bush.

Figure C-13. Blackberry.

Blueberries and Huckleberries *(Vaccinium species and Gaylussacia species)*

Description: These shrubs vary in size from 1 foot to 12 feet tall. All have alternate, simple leaves. The fruits may be dark blue, black, or red with many small seeds.

Habitat and distribution: These plants prefer open, sunny areas. They are found throughout much of the North Temperate regions and at higher elevations in Central America.

Figure C-14. Blueberry.

Bracken *(Pteridium aquilinum)*

Description: This fern has coarse, compound leaves about 3 feet long that rise from a tough, wiry rhizome.

Habitat and distribution: Bracken is found throughout most of the temperate and tropical regions of the world in open sunny areas at the margins of forests and in burned or cut forests.

Edible parts: Young leaves may be boiled and eaten as greens. Limit the amount of bracken you eat as it may contain a substance that will interfere with enzymes.

Figure C-15. Bracken.

***Breadfruit** (Artocarpus incisa)*

Description: This tree may grow up to 30 feet tall. It has dark green, deeply divided leaves that are 2-1/2 feet long and 1 foot wide. The fruits are large, green, ball-like structures up to 1 foot across when mature.

Habitat and distribution: Look for this tree at the margins of forests and homesites in the humid tropics. It is native to the South Pacific region but has been widely planted in the West Indies and parts of Polynesia.

Edible parts: The fruit pulp is edible raw. The fruit can be sliced, dried, and ground into flour for later used. The seeds are edible cooked.

Other uses: The thick sap can be used for glue and caulking material. It can also be used for birdlime (entrap small birds by smearing the sap on twigs where the birds usually perch).

Figure C-16. Breadfruit.

British soldier's lichen *(Cladonia rangiferina)*

Description: This is a low-growing, grey-green plant only a few inches tall. It does not flower but does produce bright red reproductive structures.

Habitat and distribution: Look for this lichen in open dry areas. It is very common in much of North America.

Edible parts: The entire plant is edible but has a crunchy brittle texture. Soak the plant in water with some wood ashes to remove bitterness, then dry, crush, and add to milk or to other food.

Figure C-17. British soldier's lichen.

Broad leaf lawn plantain *(Plantago major, P. rugelii)*

Description:These plants have broad leaves, over 1 inch across, that are borne close to the ground. The flowers are on a spike that arises from the middle of the cluster of leaves.

Habitat and distribution: Look for these plants in lawns and along roads in North Temperate regions.

Edible parts: The young leaves are edible raw or boiled.

Other uses: To relieve pain from wounds and sores, wash and soak the entire plant for a short while and apply it to the injured area. To treat diarrhea, drink tea made from 1 ounce of the plant boiled in 1 pint of water.

Figure C-18. Broad leaf plantain.

Buri palm *(Corypha elata)*

Description: This tree may reach 60 feet in height. It has large fan-shaped leaves up to 10 feet in length and split into about 100 narrow segments. Flowers are borne in huge clusters at the top of the tree. Following flowering, the tree dies.

Habitat and distribution: This tree grows in coastal areas of the East Indies.

Edible parts: The trunk contains starch, which is edible raw. The very tip of the trunk is also edible raw or cooked. Large quantities of liquid can be obtained by bruising the flowering stalk. The kernels of the nuts are edible.

CAUTION: The seed covering may cause dermatitis in some individuals.

Other uses: The leaves can be used for weaving.

Figure C-19. Buri palm.

Cattail *(Typha latifolia)*

Description: Cattails are grasslike plants with strap-shaped leaves ½ inch to 2 inches wide and growing up to 6 feet tall. There are several species, but all are recognized as cattails. The male flowers are borne in a dense mass above the female flowers. These last only a short time, leaving the female flowers that develop into the brown cattail. Pollen from the male flowers is often abundant and bright yellow.

Habitat and distribution: Cattails are found throughout most of the world. Look for them in full sun at the margins of lakes, streams, canals, rivers, and brackish water.

Edible parts: The young tender shoots are edible raw or cooked. The rhizome is often very tough but is a rich source of starch. Pound the rhizome to remove the starch and use as a flour. The pollen is also an abundant source of starch. When the cattail is immature and still green, the female portion may be boiled and eaten like corn on the cob.

Other uses: The dried leaves are an excellent source of weaving material and can be used to make floats and rafts. The cottony seeds make good pillow stuffing and insulation. The pollen makes excellent tinder.

Figure C-20. Cattail.

Cereus cactus *(Cereus species)*

Description: These cacti are tall and narrow with angled stems and numerous spines.

Habitat and distribution: They may be found in true deserts and other dry, open, sunny areas throughout the Caribbean region, Central America, and western United States.

Edible parts: The fruits are edible, but some may have a laxative effect.

Other uses: The pulp of the cactus is a good source of water. Break open the stem and scoop out the pulp.

Figure C-21. Cereus cactus.

Chickory *(Cichorium intybus)*

Description: This plant grows up to 6 feet tall. It has leaves clustered at the base of the stem and some leaves on the stem. The base leaves resemble those of the dandelion. The flowers are sky blue but remain open only on sunny days. Chickory has milky juice.

Habitat and distribution: Look for chickory in old fields, waste places, weedy lots, and along roads. It is a native of Europe and Asia but is found in Africa and most of North America where it grows as a weed.

Edible parts: All parts are edible. Eat the young leaves raw as a salad or boil to eat as a vegetable. Cook the roots as a vegetable. For a coffee substitute, roast the roots until they are dark brown and then pulverize.

Figure C-22. Chickory.

Coconut *(Cocos nucifera)*

Description: This tree has a single, narrow, tall trunk with a cluster of very large leaves at the top. Each leave may be over 20 feet long with over 100 pairs of leaflets.

Habitat and distribution: Coconut palms are found throughout the tropics. They are most abundant near coastal regions.

Edible parts: The nut is a valuable source of food. The milk of the young coconut is rich in sugar and vitamins and is an excellent source of liquid. The nut meat is also nutritious, but is rich in oil. To preserve the meat, spread it in the sun until completely dry.

Other uses: Use coconut oil for cooking; for protecting metal objects from corrosion; for treating saltwater sores, sunburn, and dry skin; and for improvising torches. Use the tree trunk for building material and the leaves for thatching. Hollow out the large stump to use as a food container. The coconut husks are good for flotation and the husk fibers for weaving ropes and other items. Use the gauzelike fibers at the leaf bases as strainers or use them to weave a bug net or to make a pad to use on wounds. Husk makes a good abrasive. Dried husk fiber is an excellent tinder. A smoldering husk helps to repel mosquitoes. Smoke caused by dripping coconut oil in a fire also repels mosquitoes. To render coconut oil, put the coconut meat in the sun, heat it over a slow fire, or boil it in a pot of water. Coconuts washed out to sea are a good source of freshliquid for the sea survivor.

Figure C-23. Coconut.

Cranberry *(Vaccinium macrocarpon)*

Description: This plant has tiny leaves arranged alternately. The stem creeps along the ground. The fruits are red berries.

Habitat and distribution: It only grows in open, sunny, wet areas in the colder regions of the Northern Hemisphere.

Edible parts: The berries are very tart when eaten raw. Cook in a small amount of water and add sugar if available to make a jelly.

Other uses: Cranberries may act as a diuretic.

Figure C-24. Cranberry.

Crowberry *(Empetrum nigrum)*

Description: This is a dwarf evergreen shrub with short needlelike leaves. It has small, shiny black berries that remain on the bush throughout the winter.

Habitat and distribution: Look for this plant in tundra throughout arctic regions of North America and Eurasia.

Edible parts: The fruits are edible fresh or can be dried for later use.

Figure C-25. Cowberry.

Dandelion *(Taraxacum officinal)*

Description: The leaves have a jagged edge, grow close to the ground, and are seldom more than 8 inches long. The flowers are bright yellow. There are several species of dandelions.

Habitat and distribution: Dandelions grow in open sunny locations throughout the Northern Hemisphere.

Edible parts: All parts are edible. Eat the leaves raw or cooked. Boil the roots. Roots roasted and ground are a good coffee substitute.

Other uses: The white juice in the flower stems can be used as glue.

Figure C-26. Dandelion.

Date palm *(Phoenix dactylifera)*

Description: This is a tall, unbranched tree with a crown of huge compound leaves. Its fruit is yellow when ripe.

Habitat and distribution: This tree grows in arid semitropical regions. It is native to North Africa and Arabia but has been planted in the arid semitropics in other parts of the world.

Edible parts: The fruits are edible fresh, but are very bitter if eaten before they are ripe. The fruits can be dried in the sun and preserved for a long period of time.

Other uses: The trunks provide valuable building material in desert regions where few other treelike plants are found. The leaves are durable and can be used for thatching and weaving. The base of the leaves resembles coarse cloth and can be used for scrubbing and cleaning.

Figure C-27. Date palm.

Elderberry *(Sambucus canadensis)*

Description: This is a many-stemmed shrub with opposite, compound leaves. It grows to a height of 20 feet. Its flowers are fragrant, white, and borne in large flat-topped clusters up to a foot across. Its berrylike fruits are dark blue or black when ripe.

Habitat and distribution: This plant is found in open, usually wet, areas at the margins of marshes, rivers, ditches, and lakes. It grows throughout much of eastern North America and Canada.

Edible parts: The flowers and fruits are edible. A drink can be made by soaking the flower heads for 8 hours, discarding the flowers, and drinking the liquid.

CAUTION: All other parts of the plant are poisonous and dangerous if ingested.

Figure C-28. Elderberry.

Wild figs *(Ficus species)*

Description: These trees have alternate, simple leaves with entire margins. Often, the leaves are dark green and shiny. All figs have a milky, sticky juice. The fruits vary in size depending on the species, but are usually yellow-brown when ripe.

Habitat and distribution: Figs are plants of the tropics and semitropics. They grow in several different habitats, including dense forests, margins of forests, and around human settlements.

Edible parts: The fruits are edible raw or cooked. Some figs have little flavor.

Figure C-29. Figs.

Fireweed *(Epilobium angustifolium)*

Description: This plant grows up to 6 feet tall. It has large, showy, pink flowers and lance-shaped leaves. Its relative, dwarf fireweed (Epilobium latifolium), grows 1 to 2 feet tall.

Habitat and distribution: Tall fireweed is found in open woods, on hillsides, on stream banks, and near seashores in arctic regions. It is especially abundant in burned-over areas. Dwarf fireweed is found along streams, sandbars, and lakeshores and on alpine and arctic slopes.

Edible parts: The leaves, stems, and flowers are edible in the spring but become tough in summer. The stems of old plants can be split open and the pith eaten raw.

Figure C-30. Fireweed.

Foxtail grasses *(Setaria species)*

Description: These are weedy grasses readily recognized by the narrow, cylindrical head containing long hairs. The grains are small, less than 1/4-inch long. The dense heads of grain often droop when ripe.

Habitat and distribution: Look for foxtail grasses in open sunny areas, along roads, and at the margins of fields. Some species occur in wet marshy areas. Species of *Setaria* are found throughout the United States, Europe, western Asia, and tropical Africa. In some parts of the world foxtail grasses are grown as a food crop.

Edible parts: The grains are edible raw, but are very hard and sometimes bitter. Boiling removes some of the bitterness and makes them easier to eat.

Figure C-31. Foxtail.

Iceland moss *(Cetraria islandica)*

Description: This moss grows only a few inches high; its color may be grey, white, or even reddish.

Habitat and distribution: Look for it in open areas. It is found only in the arctic.

Edible parts: All parts of the Iceland moss are edible. During the winter or dry season, it is dry and crunchy but softens when soaked. Boil the moss to remove bitterness. After boiling, eat or add to milk or grains as a thickening agent. Dried plants store well.

Figure C-32. Iceland moss.

Indian almond *(terminalia catappa)*

Description: This tree grows up to 30 feet tall. Its leaves are evergreen, leathery, 18 inches long, 6 inches wide, and very shiny. It has small, yellowish-green flowers. The fruits are flat, 4 inches long, and not quite as wide. They are green when ripe.

Habitat and distribution: This tree is usually found growing near the ocean. It is a common and often abundant tree in the Caribbean and Central and South America. It can also be found in tropical rain forests of southeastern Asia, northern Australia, and Polynesia.

Edible parts: The seed is a good source of food. Remove the fleshy green covering and eat the seed raw or cooked.

Figure C-33. Indian almond.

Indian potato (Eskimo potato) *(Claytonia species)*

Description: All species of Claytonia are somewhat fleshy plants only a few inches tall with showy flowers about an inch across.

Habitat and distribution: Some species occur in rich forests where they are conspicuous before the leaves develop. Western species are found throughout most of northern United States and in Canada.

Edible parts: The tubers are edible but should be boiled before eating.

Figure C-34. Indian potato.

Juniper *(Juniperus species)*

Description: Junipers, sometimes called cedars, are trees or shrubs with very small scalelike leaves densely crowded around the branches. Each leaf is less than ½ inch long. All species have a distinct aroma resembling the well known cedar. The berrylike cones are usually blue and covered with a whitish wax.

Habitat and distribution: Look for junipers in open, dry, sunny areas throughout North America and northern Europe. Some species are found in southeastern Europe, across Asia to Japan, and in the mountains of North Africa.

Edible parts: The berries and twigs are edible. Eat the berries raw or roast the seeds to use as a coffee substitute. Use dried and crushed berries as a seasoning for meat. Gather young twigs to make a tea.

CAUTION: There are many plants that may be called cedars but be no relation to junipers and may be harmful. Always look for the berrylike structures, needle leaves, and resinous, fragrant sap to be sure the plant you have is a juniper.

Figure C-35. Juniper.

Lotus *(Nelumbo)*

Description: There are two species of lotus; one has yellow flowers and the other pink flowers. The flowers are large and showy. The leaves, which may float on or rise above the surface of the water, often reach 5 feet in radius. The fruit has a distinctive flattened shape and contains up to 20 hard seeds.

Habitat and distribution: The yellow-flowered lotus is native to North America. The pink-flowered species, which is widespread in the Orient, is planted in many other areas of the world. Lotuses are found in quiet freshwater.

Edible parts: All parts of the plant are edible raw or cooked. The underwater parts contain large quantities of starch. Dig the fleshy portions from the mud and bake or boil them. Boil the young leaves and eat as a vegetable. The seeds have a pleasant flavor and are nutritious. Eat them raw or parch and grind into flour.

Figure C-36. Lotus.

Malanga *(Xanthosoma caracu)*

Description: This plant has soft, arrow-shaped leaves, up to 2 feet long. The leaves have no above-ground stems.

Habitat and distribution: This plant grows widely in the Caribbean region. Look for it in open, sunny fields.

Edible parts: The tubers are rich in starch. Cook them before eating. This will destroy a chemical that is contained in all parts of the plant.

Figure C-37. Malanga.

Mango *(Mangifera indica)*

Description: This tree may reach 100 feet in height. It has alternate, simple, shiny, dark green leaves. The flowers are small and inconspicuous. The fruits have a large single seed. There are many cultivated varieties of mango. Some have red flesh, others yellow or orange.

Habitat and distribution: This tree grows in warm, moist regions. It is native to northern India, Burma, and West Malaysia, but is now grown throughout the tropics.

Edible parts: The fruits are a nutritious source of food. The flesh can be eaten green by shredding and eating like a salad. The ripe fruit can be peeled and eaten raw.

CAUTION: If you are sensitive to poison ivy, you should avoid eating mangos as they cause a severe reaction in sensitive individuals.

Figure C-38. Mango.

Marsh marigold *(Caltha palustris)*

Description: This plant has rounded, dark green leaves arising from a short stem. It has bright yellow flowers.

Habitat and distribution: This plant is found in bogs, lakes, and slow-moving streams. It is abundant in arctic and subarctic regions and in much of the eastern region of the northern United States.

Edible parts: All parts are edible if boiled.

CAUTION: Do not eat raw.

Figure C-39. Marsh marigold.

Mulberry *(Morus species)*

Description: This tree has alternate, simple, often lobed leaves with rough surfaces. The fruits are blue or black and many seeded.

Habitat and distribution: Mulberry trees are found in forests, along roadsides, and in abandoned fields in temperate and tropical zones of North America, South America, Europe, Asia, and Africa.

Edible parts: The fruit is edible raw or cooked. It can be dried for eating later.

CAUTION: When eaten in quantity, it acts as a laxative.

Other uses: The inner bark of the tree can be shredded and used to make twine or cord.

Figure C-40. Mulberry.

Narrow leaf lawn plantain *(Plantago lanceolata)*

Description: This plant has leaves up to 5 inches long and 1 inch wide covered with hairs. The leaves form a rosette. The flowers are small and inconspicuous.

Habitat and distribution: This plant is a common weed throughout much of North American and Europe.

Edible parts: The young, tender leaves are edible raw. Older leaves should be cooked.

Figure C-41. Narrow leaf plantain.

Nettle *(Urtica and Laportea species)*

Description: These plants grow several feet high. They have small, inconspicuous flowers. Fine hairlike bristles cover the stems, leafstalks, and undersides of leaves. The bristles cause a stinging sensation when they touch the skin.

Habitat and distribution: Nettles prefer moist areas along streams or at the margins of forests. They are found throughout North America, Central America, the Caribbean, and northern Europe.

Edible parts: Young shoots and young leaves are edible. Boiling the plant destroys the stinging element of the bristles.

Other uses: Mature stems have a fibrous layer that can be divided into individual fibers and used to weave string or twine.

Figure C-42. Nettle.

Nipa palm *(Nipa fruticans)*

Description: This palm has a short trunk, which is mainly underground, and very large, erect leaves up to 20 feet tall. The leaves are divided into leaflets. A flowering head forms on a short erect stem that rises among the palm leaves. The fruiting (seed) head is dark brown and may be a foot in diameter.

Habitat and distribution: This palm is common on muddy shores in coastal regions throughout eastern Asia.

Edible parts: The young flower stalk and the seeds provide a good source of water and food. Cut the flower stalk and collect the juice. The juice is rich in sugar. The seeds are hard, but edible.

Other uses: The leaves are excellent for thatching and coarse weaving.

Figure C-43. Nipa palm.

Nodding onion *(Allium cernuum)*

Description: Allium cernuum is an example of the many species of wild onions and garlics, all of which are easily recognized by their distinctive odor.

Habitat and distribution: Wild onions and garlics can be found in open sunny areas throughout the temperate regions. Cultivated varieties may be found anywhere in the world.

Edible parts: The bulbs and young leaves are edible.

CAUTION: There are several plants with onionlike bulbs that are extremely poisonous. Be certain that the plant you are using is a true onion or garlic.

Other uses: Eating large quantities of onions will give your body an odor that will help to repel insects.

Figure C-44. Nodding onion.

Nutsedge *(Cyperus esculentus)*

Description: This very common plant has a triangular stem and grasslike leaves. It grows to a height of 8 to 24 inches. The mature plant has a soft burrlike bloom that extends from a whorl of leaves. Tubers ½ to 1 inch in diameter grow at the ends of the roots.

Habitat and distribution: Nutsedge grows in moist sand areas throughout the world. It is often an abundant weed in cultivated fields.

Edible parts: The tubers are edible raw, boiled, or baked. They can also be ground and used as a coffee substitute.

Figure C-45. Nutsedge.

Oaks *(Quercus species)*

Description: Oak trees have alternate leaves and acorn fruits. There are two main groups of oaks: red and white. The red oak group has leaves with bristles and smooth bark in the upper part of the tree; red oak acorns take 2 years to mature. The white oak group has leaves without bristles and rough bark in the upper portion of the tree; white oak acorns mature in 1 year.

Habitat and distribution: Oak trees are found in many habitats throughout North America, Central America, and parts of Europe and Asia.

Edible parts: All parts are edible, but often contain large quantities of bitter substances. White oak acorns generally have a better flavor than the red oak acorns. Gather and shell the acorns. Soak red oak acorns in water for 1 to 2 days to remove the bitter substance. You can speed up this process by putting wood ashes in the water in which you soak the acorns. Boil the acorns or grind them into flour and use the flour for baking. Acorns baked until very dark can be used as a coffee substitute.

Other uses: Oak wood is excellent for building or burning. Small oaks can be split and cut into long thin strips 1/8 to 1/4-inch thick and 1/2-inch wide and used for weaving mats, baskets, or frameworks for packs, sleds, furniture, and so forth. Oak bark soaked in water produces a tanning solution for preserving leather.

Figure C-46. Oak.

Orach *(Atriplex species)*

Description: This plant is vinelike in growth and has arrowhead-shaped leaves up to 2 inches long.

Habitat and distribution: Orach species are entirely restricted to salty soils. They are found along North America coasts and on the shores of alkaline lakes inland. They are also found along seashores from the Mediterranean countries to inland areas in North Africa and eastward to Asia Minor and central Siberia.

Edible parts: The plant is edible raw or boiled.

Figure C-47. Orach.

Palmetto palm *(Sapal palmetto)*

Description: These are tall unbranched trees with persistent leaf bases for most of the trunk. The leaves are large, simple, and palmately lobed. Fruits are dark blue or black with a hard seed.

Habitat and distribution: The palmetto palm is found throughout the coastal regions of southeastern United States.

Edible parts: The fruits are edible raw. The hard seeds may be ground into flour. The heart of the palm is a nutritious food source at any season. Cut off the top of the tree to obtain the palm heart.

Figure C-48. Palmetto palm.

Persimmon *(Diospyros virginiana and other species)*

Description: These trees have alternate, dark green, elliptic-shaped leaves with entire margins. The flowers are inconspicuous. The fruits are orange, have a sticky consistency, and have several seeds.

Habitat and distribution: The persimmon is a common forest margin tree. It is widespread in Africa, eastern North America, and the Far East.

Edible parts: The leaves are a good source of vitamin C. The fruits are edible raw or baked. To make tea, dry the leaves and soak them in hot water.

CAUTION: Some persons are unable to digest persimmon pulp.

Figure C-49. Persimmon.

Pigweed *(Amaranthus)*

Description: These plants, which grow 3 to 5 feet tall, are abundant weeds in many parts of the world. All amaranthus have alternate simple leaves; they may have some red color present on the stems. They bear minute flowers in dense clusters at the top of the plants. Seeds may be brown or black.

Habitat and distribution: Look for amaranthus along roadsides, in disturbed waste areas, or as weeds in crops throughout the world. Some species of amaranthus have been grown as a grain crop in South America.

Edible parts: All parts are edible, but some may have sharp spines that should be removed before eating. The young plants or growing tips of older plants are an excellent vegetable. Simply boil the young plants or eat them raw. The seeds are very nutritious. Shake the tops of older plants to get the seeds. Eat the seeds raw, boiled, or ground into flour.

Figure C-50. Pigweed.

Pincushion cactus *(Mammalaria species)*

Description: Members of this cactus group are round, short, and barrel-shaped without leaves. The entire plant is covered with sharp spines.

Habitat and distribution: These cacti are found throughout much of the desert regions of western United States and parts of Central America.

Use: They are a good source of water in the desert.

Figure C-51. Pincushion cactus.

Pines (Pinus species)

Description: Pine trees are easily recognized by their needlelike leaves grouped in bundles. Each bundle may contain one to five needles, the number varying among species. The odor and sticky sap of the tree provide a simple way to distinguish pines from similar looking trees with needlelike leaves.

Habitat and distribution: Pines prefer open, sunny areas. They are found throughout North America, Central America and much of the Caribbean region, North Africa, the Middle East, Europe, and some places in Asia.

Edible parts: The seeds of all species are edible. The young male cones, borne only in the spring of the year, can be collected as a survival food. The bark of young twigs is edible. Boil or bake the young cones. Peel off the bark of thin twigs. The juicy inner bark can be chewed; it is rich in sugar and vitamins, especially in the spring when the sap is rising.

Other uses: The resin can be used to waterproof articles. It can also be used as glue. Collect the resin from the tree. If there is not enough on the tree, cut a V notch in the bark so more sap will seep out. Put the resin in a container and heat it. The hot resin is your glue. Use it as is, or add a small amount of ash dust to strengthen it. Use it immediately.

Figure C-52. Pine.

Pokeweed *(Phytolacca americana)*

Description: This plant may grow as high as 10 feet. Its leaves are elliptic and up to 3 feet in length. Pokeweed produces many large clusters of purple fruits late in the season.

Habitat and distribution: Look for this plant in open, sunny areas in forest clearings, in fields, and along roadsides in eastern North America, Central America, and the Caribbean.

Edible parts: The young leaves and stems are edible cooked; boil them twice, discarding the water from the first boiling. The fruits are edible if cooked.

CAUTION: All parts of this plant are poisonous if eaten raw. Never eat the underground portions of the plant as these contain the highest concentrations of the poisons.

Other uses: Use the juice of fresh berries as a dye.

Figure C-53. Pokeweed.

Prickly pear cactus *(Opuntia)*

Description: This cactus has flat, padlike stems that are green and covered with many round furry dots that contain sharp-pointed hairs.

Habitat and distribution: It is found throughout most of the United States and Central and South America in arid and semiarid regions and in dry sandy areas of wetter regions. Some species are planted in arid and semiarid regions of other parts of the world.

Edible parts: All parts of the plant are edible. Peel the fruits and eat fresh or crush to prepare a refreshing drink. Take care to avoid the tiny pointed hairs.

Other uses: The pads are a good source of water. Peel them carefully to remove all sharp hairs before putting in your mouth.

The pads can also be used to promote healing. Split them and apply the pulp to wounds.

Figure C-54. Prickley pear cactus.

Purslane *(Portulaca oleracea)*

Description: This plant grows close to the ground. It is seldom more than a few inches tall. Its stems and leaves are fleshy and often tinged with red. It has paddle-shaped leaves, 1 inch or less long, clustered at the tips of the stems. The flowers are yellow or pink. The seeds are tiny and black.

Habitat and distribution: It grows in full sun in cultivated fields, field margins, and other weedy areas throughout the world.

Edible parts: All parts are edible. Wash and boil the plants for a tasty vegetable or eat the plants raw. Use the seeds as a flour substitute or eat them raw.

Figure C-55. Purslane.

Reed *(Phragmites australis)*

Description: This tall coarse grass grows to 12 feet tall and has grey-green leaves about 1-1/2 inches wide. It has large masses of brown flower branches in early summer. These rarely produce grain. By late in the season these become fluffy gray masses.

Habitat and distribution: Look for reed in any open, wet area, especially one that has been disturbed through dredging. Reed is found throughout the temperate regions of both the Northern and Southern Hemispheres.

Edible parts: All parts of the plant are edible raw or cooked at any season. Harvest the stems as they emerge from the soil and boil them. Or harvest them just before they produce flowers, then dry and beat them into flour. The underground stems can be dug and boiled, but they are often tough. Seeds can be eaten raw or boiled, but they are rarely found.

Figure C-56. Reed.

Rock polypody *(Polypodium virginianum)*

Description: This fern and related species have small leaves about 5 inches long and 2 inches wide and a hairy, creeping rhizome. Some of the leaves will have small round dots on their undersurface.

Habitat and distribution: This fern is common on rocky, shady slopes. It and related species are widespread in eastern North and Central America.

Edible parts: The young leaves are edible.

Figure C-57. Rock polypody.

Rock tripe *(Umbelicaria species)*

Description: This plant forms large patches with curling edges. The top of the plant is usually black. The underside is lighter in color.

Habitat and distribution: Look on rocks and boulders for this plant. It is common throughout North America.

Edible parts: The entire plant is edible. Scrape it off the rock and wash it to remove grit. The plant may be dry and crunchy; soak it in water until soft. Rock tripes may contain large quantities of bitter substances; soaking or boiling in several changes of water will remove the bitterness.

CAUTION: There are some reports of poisoning from rock tripe, so apply the Universal Edibility Test.

Figure C-58. Rock tripe.

Rose apple *(Eugenia jambos)*

Description: This tree grows 10 to 30 feet high. It has opposite, simple, dark green, shiny leaves. It has fluffy, yellowish-green flowers and egg-shaped fruit, red to purple when fresh.

Habitat and distribution: This tree is widely planted in all of the tropics, where it may also be found in the semiwild state. It usually occurs in thickets, waste places, and secondary forests.

Edible parts: The fruit is edible raw or cooked.

Figure C-59. Rose apple.

St. John's bread *(Ceratonia siliqua)*

Description: This is a large tree with a spreading crown. Its leaves are compound and alternate. Its seed pods are up to 1½ feet long and are filled with round, hard seeds and a thick pulp.

Habitat and distribution: This tree is found throughout the Mediterranean, the Middle East, and parts of North Africa.

Edible parts: The young tender pods are edible raw or boiled. The seeds produced in mature pods can be pulverized and cooked as porridge.

Figure C-60. St. John's bread.

Sassafras *(Sassafras albidum)*

Description: This shrub or small tree has different leaves borne on the same plant. Some leaves will have one lobe, some two lobes, and some no lobes. The flowers, which appear in early spring, are small and yellow. The fruits are dark blue.

Habitat and distribution: Sassafras grows at the margins of roads and forests, usually in open sunny areas. It is a common tree throughout Eastern North America.

Edible parts: The young twigs and leaves can be eaten fresh or dried. The dried young twigs and leaves can be added to soups. Dig the underground portion, peel off the bark, and let it dry. Then boil it in water to prepare sassafras tea.

Figure C-61. Sassafras.

Sorghum *(Sorghum species)*

Description: There are many different kinds of sorghum, all of which have grains borne in heads at the top of the plants. The grains are brown, white, red, or black. Sorghum is the main food crop in many parts of the world.

Habitat and distribution: Sorghum is found worldwide, usually in warmer climates. All species occur in open, sunny areas.

Edible parts: The grains are edible at any stage of development. When young, the grains are milky and can be eaten raw. When older, the grains should be boiled. Sorghum is a nutritious food.

Other uses: The stems of tall sorghums can be used for building.

Figure C-62. Sorghum.

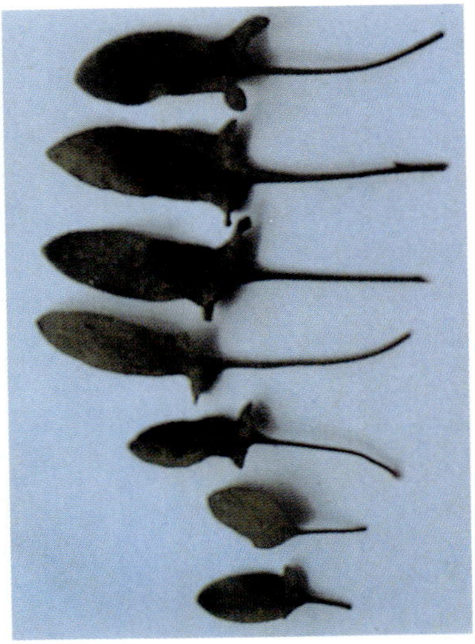

Sorrel *(Rumex acetosella)*

Description: These plants are seldom more than a foot tall. They have alternate leaves often with arrowlike bases, very small flowers, and frequently reddish stems.

Habitat and distribution: Look for these plants in old fields and other disturbed areas in North America and Europe.

Edible parts: The plants are edible raw or cooked.

CAUTION: These plants contain oxalates that can be damaging if too many plants are eaten raw. Cooking seems to destroy the oxalates.

Figure C-63. Sorrel.

Spatterdock, yellow water lily *(Nuphar)*

Description: This plant has leaves up to 2 feet long with a triangular notch at the base; the shape of the leaves is somewhat variable. The yellow flowers are 1 inch across and develop into bottle-shaped fruits. The fruits are green when ripe.

Habitat and distribution: These plants grow throughout most of North America. They are found in quiet, fresh, shallow water (never deeper than 6 feet).

Edible parts: All parts of the plant are edible. The fruits contain several dark brown seeds that can be parched or roasted then ground into flour. The large rootstock is filled with starch. Dig it out of the mud, peel off the outside, and boil the flesh. Sometimes the rootstock contains large quantities of a very bitter compound. Boiling in several changes of water may remove this.

Figure C-64. Spatterdock, yellow water lily.

Sugar palm *(Arenga pinnata)*

Description: This tree grows about 50 feet high and has huge leaves up to 20 feet long. Needlelike structures stick out of the bases of the leaves. Flowers grow below the leaves and form large conspicuous clusters from which the fruits arise.

Habitat and distribution: This palm is native to the East Indies but has been planted in many parts of the tropics. It can be found at the margins of forests.

Edible parts: The chief use of this palm is for sugar. However, the seeds and the tip of the stem may be used as survival food. Bruise a young flower stalk with a stone or similar object; collect the juice as it comes out. It is an excellent source of sugar. Boil the seeds. Use the tip of the stem as a vegetable.

CAUTION: The flesh covering the seeds may cause dermatitis.

Other uses: The shaggy material at the base of the leaves makes an excellent rope as it is strong and resists decay.

Figure C-65. Sugar palm.

Sugarcane *(Saccharum officinarum)*

Description: This plant grows up to 15 feet tall. It is a grass and has grasslike leaves. The stems, which are green or reddish, are swollen where the leaves arise. Cultivated sugarcane seldom flowers.

Habitat and distribution: Look for sugarcane in fields. It grows only in tropical areas, but is found in the tropics throughout the world. Because it is a crop, it is often found in large numbers.

Edible parts: The stem is an excellent source of sugar and is very nutritious. Peel the outer portion off with your teeth and eat the sugarcane raw. You can also squeeze juice out of the sugarcane.

Figure C-66. Sugarcane.

Sweetsop *(Annona squamosa)*

Description: This tree is small, seldom more than 20 feet tall, and much branched. It has alternate, simple, elongate leaves that are dark green in color. The fruit is green when ripe, round in shape, and covered with protruding bumps on the surface. The flesh is white and creamy.

Habitat and distribution: Look for sweetsop at margins of fields, near villages, and around homesites in tropical regions.

Edible parts: The fruit flesh is edible raw.

Other uses: The seeds, when ground fine, can be used as an insecticide.

CAUTION: The ground seeds are extremely dangerous to the eyes.

Figure C-67. Sweetsop.

Taro, cocoyam, elephant ears, eddo, dasheen *(Colocasia and Alocasia species)*

Description: All plants in these groups have large leaves, sometimes up to 6 feet tall, that arise from a very short stem. The rootstock is thick and fleshy and filled with starch.

Habitat and distribution: These plants grow in the humid tropics. Look for them in fields and near homesites and villages.

Edible parts: All parts of the plant are edible when boiled.

CAUTION: If eaten raw, these plants will cause a serious inflammation of the mouth and throat.

Figure C-68. Taro.

Ti (Cordyline terminalis)

Description: The ti plant has unbranched stems with straplike leaves often clustered at the tip of the stem. The leaves vary in color and may be green or reddish. Flowers are borne at the apex of the plant in large plumelike clusters. The ti plant may grow up to 15 feet tall.

Habitat and distribution: Look for this plant at the margins of forests or near homesites in tropical areas. It is native to the Far East but is now widely planted in tropical areas worldwide.

Edible parts: The roots and very young leaves are good survival food. Boil or bake the short, stout roots at the base of the plant. They are a valuable source of starch. Boil the very young leaves to eat. Leaves may be used to wrap other food to cook over coals or to steam.

Other uses: Use the leaves to cover shelters or to make a rain cloak. Cut the leaves into liners for shoes; this works especially well if you have a blister. Fashion temporary sandles from the ti leaves. The terminal leaf, if not completely unfurled, can be used as a sterile bandage.

Cut the leaves into strips, then braid or weave the strips into rope. Make a sling-type basket from six leaves: Split the tip of each leaf in several places and tie the tips of two leaves together. Fit the three pairs of tied leaves together to form the basket.

Figure C-69. Ti plant.

Tree ferns *(various genera)*

Description: These are tall trees with long slender trunks that often have a very rough barklike covering. Large lacy leaves uncoil from the top of the trunk.

Habitat and distribution: Tree ferns are found in wet tropical forests.

Edible parts: The young leaves and the soft inner portion of the trunk are edible. Boil the young leaves and eat as a green. Eat the inner portion of the trunk raw or bake it.

Figure C-70. Tree fern.

Water chestnut *(Trapa natans)*

Description: This is an aquatic plant that roots in the mud and has underwater leaves that are finely divided. The floating leaves are much larger and coarsely toothed. The fruits, borne underwater, have four sharp spines on them.

Habitat and distribution: Look for these plants in freshwater only. They are a native of Asia, but have spread to many parts of the world in both temperate and tropical areas.

Edible parts: The fruits can be eaten raw or cooked. The seeds are also a source of food.

Figure C-71. Water chestnut.

Water fern *(Ceratopteris thalictroides)*

Description: These are spongy, soft ferns 2 to 3 feet tall with smooth leaves.

Habitat and distribution: They are found throughout the tropics, floating in water or growing at the edge of small streams and lakes.

Edible parts: The entire plant is edible when boiled.

Figure C-72. Water fern.

Water lily *(Nymphaea odorata)*

Description: These plants have large triangular leaves that float on the surface of the water, large flowers that are usually white or red and fragrant, and thick, fleshy rhizomes that grow in the mud.

Habitat and distribution: Water lilies are found throughout much of the temperate and subtropical regions.

Edible parts: The flowers, seeds, and rhizomes are edible raw or cooked. To prepare rhizomes for eating, peel off the corky rind. Eat raw, or slice thinly, allow to dry, and then grind into flour. Dry, parch, and grind the seeds into flour.

Other uses: The liquid that results from boiling the thickened root in water is recommended as a medicine for diarrhea and as a gargle for sore throats.

Figure C-73. Water lily.

Water plantain *(Alisma plantago-aquatica)*

Description: This plant has small, white flowers and heart-shaped leaves with pointed tips. The leaves are clustered at the base of the plant.

Habitat and distribution: Look for this plant in freshwater and in wet areas in full sun in temperate and tropical zones.

Edible parts: The rootstocks are a good source of starch. Boil or soak them in water to remove the bitter taste.

Figure C-74. Water plantain.

Wild caper *(Capparis aphylla)*

Description: This is a thorny shrub that loses its leaves during the dry season. Its stems are grey green in color and its flowers are pink.

Habitat and distribution: These shrubs form large stands in scrub and thorn forests and in desert scrub and waste. They are common throughout North Africa and the Middle East.

Edible parts: The fruits and the buds of young shoots are edible raw.

Figure C-75. Wild caper.

Wild rice *(Zizania aquatica)*

Description: Wild rice is a tall grass that averages 3 to 5 feet in height, but may reach 15 feet. The grain is borne in very loose heads at the top of the plant. It is dark brown or blackish when ripe.

Habitat and distribution: Wild rice grows only in very wet areas in tropical and temperate regions.

Edible parts: During the spring and summer, the central portion of the lower stems and root shoots are edible. Remove the tough covering before eating. During the late summer and fall, collect the straw-covered husks. Dry and parch the husks, break them, and remove the rice. Boil the rice, or roast it and then beat it into flour.

Figure C-76. Wild rice.

Wild rose *(Rosa species)*

Description: This shrub grows 2 to 8 feet high. It has alternate leaves and sharp prickles. The flowers may be red, pink, or yellow. The fruits, called rose hips, remain on the shrub year round.

Habitat and distribution: Look for wild roses in dry fields and open woods throughout the Northern Hemisphere.

Edible parts: The flowers and buds are edible raw or boiled. In an emergency, the young shoots can be peeled and eaten. Fresh, young leaves can be boiled in water to make tea. After the flower petals fall, eat the rose hips; the pulp is highly nutritious and an excellent source of vitamin C. Crush or grind dried rose hips to make flour.

CAUTION: Eat only the outer portion of the fruit as the seeds of some species are quite prickly and can cause internal distress.

Figure C-77. Wild rose.

Yam *(Dioscorea species)*

Description: These plants are vines that creep along the ground. They have alternate, heart- or arrow-shaped leaves. Their rootstock may be very large and weigh many pounds.

Habitat and distribution: True yams are restricted to tropical regions where they are an important food crop. Look for yams in fields.

Edible parts: Boil the rootstock and eat it as a vegetable.

Figure C-78. Yam.

Common and scientific name	Description	Location
Almond *(Prunus amygdalus)*	A mature almond tree looks like a peach tree and sometimes grows 40 feet tall. The fruits resemble gnarled unripe peaches and grow in clusters. The mature fruit splits open down the side, exposing the ripe almond nut. This tree is found in tropical scrub and thorn forests, in temperate zone evergreen scrub forests, and in desert scrub and waste areas of southern Europe, the eastern Mediterranean countries, Iran, Arabia, China, Madeira, the Azores, and the Canary Islands. Crack the almond nut to get to the edible kernel.	*Tropics* *Temperate*
Arctic willow *(Salix arctica)*	This small shrub grows 1 to 2 feet high in clumps that form dense mats on the tundra. This shrub is common in all tundra areas in North America, Europe, and Asia. The young leaves and stems are tasty when cooked.	*Arctic*
Bael fruit *(Aegle marmelos)*	This tree is 8 to 15 feet tall with a dense, spiny growth. It has elliptic-shaped, toothed leaves and a greenish white flower when in bloom. The fruit is 2 to 4 inches in diameter, gray or yellowish, and full of seeds. This tree is found in tropical rain forests and semievergreen seasonal forests in central and southern India bordering the Himalaya mountains and in Burma. The fresh fruit is sour but refreshing.	*Tropics*
Bistort (Knotweed) *(Polygonum)*	This plant has small white or pink flowers that form on slender spikes. It has long pointed, toothless leaves that grow from the stem near the soil level. It grows in dry tundra. The root is rich in starch but slightly acid tasting if eaten raw. Soak the root several hours in water and then roast it.	*Arctic*

Figure C-79. Edible wild plants.

Common and scientific name	Description	Location
Burdock *(Arctium lappa)*	This plant has wavy-edged leaves and flower heads in bristly receptacles. It is found growing in open wasteland during spring and summer. Peel the tender leaf stalks and eat them raw or cook them like greens. The roots are also edible. Other uses: A liquid made from the roots will help to produce sweating and increase urination. Dry the root, simmer it in water, strain the liquid, and then drink the strained liquid. Fiber from the dried stock can be used to weave cordage.	*Temperate*
Cashew nut *(Anacardium accidentale)*	This spreading evergreen tree grows up to 40 feet high. It has toothless leaves 8 inches long and 4 inches wide, yellowish-pink flowers, and thick, pear-shaped fruit that may be red or yellow when ripe. At the tip of the fruit is a hard kidney-shaped nut, green or brown depending on its stage of maturity. This tree is found in tropical rain forests, semievergreen seasonal forests, and scrub and thom forests of northern South America, West Indies, Africa, and India. The fruit is edible. The nut encloses one seed, which is edible when roasted. *CAUTION: The green hull contains a resinous poison that will blister lips and tongue. Roasting the nuts will destroy the poison.*	*Tropics*
Coltsfoot *(Petasites)*	This plant has a scaley, fleshy stalk about a foot high topped by a cluster of creamy flowers. Its leaves are thick, large, somewhat rounded, 3 to 10 inches long, dark green on top, and fuzzy white underneath. This plant is found in moist woods and wet tundra during the spring and summer. The leaves and flowering shoots are edible.	*Arctic*

Figure C-79. Edible wild plants. (Continued)

Common and scientific name	Description	Location
Goa bean *(Psophocarpus tetragonolobus)*	This climbing plant may cover small shrubs and trees. It has toothless, egg-shaped leaves 6 inches long, bright blue flowers, and bean pods 9 inches long. The mature pods are four-angled with jagged winglike edges. This plant is found in clearings and around abandoned garden sites in tropical rain forests and semievergreen seasonal forests of Africa, Asia, the East Indies, the Philippines, and Formosa. Young pods, mature seeds, thick roots, young leaves, and bean sprouts are edible. Prepare the young pods like green beans. Parch or roast the mature seeds over hot coals, or place them in damp moss to germinate and then eat the resultant bean sprout. Eat the roots raw. Steam the leaves or eat them raw.	*Tropics*
Salmonberry (Cloudberry) *(Rubus chameamorus)*	This is an erect plant that seldom grows over 1 foot in height. It has a few large, wide, five-lobed leaves. The flowers are white, and the berries, which ripen in July or August, are amber or salmon-colored. This plant grows in tundra areas and often covers many acres of ground. The berries are edible.	*Arctic*
Thistle *(Carduacea)*	This plant may grow as high as 5 feet. Its leaves are long pointed, deeply lobed, and prickly. It grows worldwide in dry woods and fields. Peel the stalks, cut them into short sections, and boil them to eat. The roots are edible raw or cooked. *(CAUTION: The juice of the thistle is a powerful laxative.)* Other uses: Twist the tough fibers of the stems to make a strong twine.	*Worldwide*
Woolly lousewort *(Pedicularis)*	This plant has thick, strong, rose-colored stems that are often visible above the snow. It grows mostly in dry tundra. The root is edible raw or cooked.	*Arctic*

Figure C-79. Edible wild plants. (Continued)

Common and Scientific Name	Description	Location
Wild desert gourd *(Citrullus colocynthis)*	This ground-trailing vine grows 8 to 10 feet long. It is a member of the watermelon family and its leaves resemble those of the watermelon. Its round gourds grow the size of an orange and are yellow in color when ripe. This plant is found in desert scrub and waste and in any climatic zone in the Sahara, in Arabia, on the southeastern coast of India, and on some islands in the Aegean Sea. It will grow in the hottest localities. The flowers are edible. The seeds inside the ripe gourd are edible roasted or boiled. Completely separate them from the bitter pulp before cooking. *CAUTION: The pulp inside the gourd is a violent purgative. Some seeds are reported to be poisonous so use the edibility test.* Other uses: The stem tips are full of water and may be chewed.	*Tropics*

Figure C-79. Edible wild plants. (Continued)

Seaweed. One plant that should never be overlooked is seaweed (figure C-81). It is a form of marine algae found on or near ocean shores. There is also one edible freshwater variety. Seaweed is a valuable source of iodine, other minerals, and vitamin C. It can, however, have a strong laxative effect if eaten in large quantities when the stomach is not conditioned to it.

In gathering seaweeds for food, find plants attached to rocks or floating free. Seaweed stranded on the beach for any length of time may be spoiled or decayed.

Preparation for eating depends on the type of seaweed. Thin, tender varieties can be dried in the sun or over a fire until crisp, then crushed and used in soup or broth. Thick, leathery seaweeds should be washed, boiled for a short time to soften, and eaten with other foods. Although some kinds can be eaten raw, first test them for edibility.

Common and Scientific Name	Description
Green seaweed *(Ulva lactuca)*	Common green seaweed, often called sea lettuce, occurs abundantly on both sides of the Pacific and North Atlantic oceans. Wash it in clean water and use it like garden lettuce.
Sugar wrack *(Laminaria saccharina)*	This seaweed is found on both sides of the Atlantic Ocean and on the coasts of China and Japan. It is brown in color.
Edible kelp *(Alaria esculenta)*	This seaweed is found in both the Atlantic and Pacific Oceans, usually below the high-tide line on submerged ledges and rocky bottoms. It has a short cylindrical stem and thin, wavy olive-green or brown fronds 1 foot to several feet in length. Boil the kelp to soften it, then mix it with vegetables or soup.
Irish moss *(Chondrus crispus)*	This seaweed is found on both sides of the Atlantic Ocean, usually just below the high-tide line. It is often cast up on the shore. It is brown and has a tough, elastic, leathery texture. When dried, it shrinks and becomes crisp. Boil it for eating.
Dulse *(Rhodymenia palmata)*	This seaweed is found on both sides of the Atlantic Ocean and on the coasts of the Mediterranean Sea. It is found attached to rocks or coarser seaweeds, usually at the low-tide level. It has a very short stem that quickly broadens to a thin, broad, fan-shaped, dark red expanse that is divided by clefts into several short, round-tipped lobes. Dulse is leathery in consistency but has a sweet taste. If dried and rolled, it can substitute for chewing tobacco.
Laver *(Porphyra)*	This seaweed is common to both the Atlantic and Pacific areas. It is usually found on the beach at the low-tide level. It is red, dark purple, purplish-brown in color and has a satiny sheen or a filmy luster. Clean the seaweed and use as a relish or boil gently until tender. Another way to prepare it for eating is to pulverize it, add to crushed grains, and fry the mixture like flatcakes. It also has thirst-quenching value when chewed.
Red warm-water seaweed	A great variety of red seaweed is found in the South Pacific area and accounts for a large portion of the native diet.
Freshwater algae *(Nostoc commune)*	A freshwater variety of seaweed common in China, America, and Europe is the marine alga known as common Nostoc. It forms green, round jellylike lobules about the size of marbles. Dry it and use it in soups.

Figure C-81. Descriptions of seaweeds.

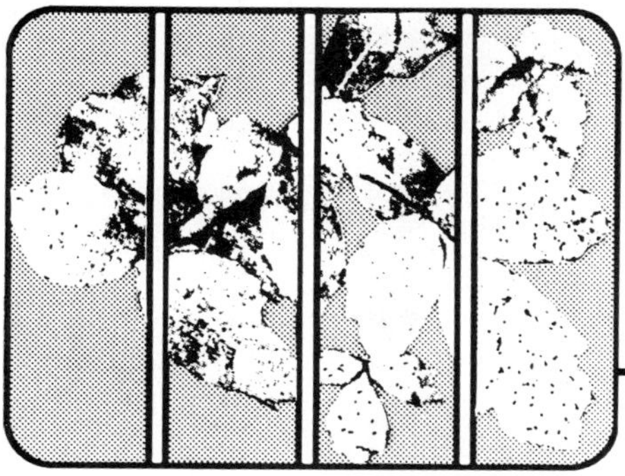

Appendix D

Poisonous Plants

There are two general types of poisonous plants—those that cause dermatitis and those that cause internal poisoning. There are also plants that if ingested cause the person to have a serious reaction when exposed to sun, and plants that have poison nectar, which produces poison honey.

External Poisoning

Avoid the plants that cause dermatitis, such as poison ivy, poison oak, and poison sumac (figures D-1 and D-2). These three plants, which are common in the United States, have counterparts in other regions of the world. If you come in contact with this type plant, wash the poisonous oil away with soap and water and apply a bandage to prevent spreading.

Internal Poisoning

Even more serious than dermatitis-causing plants are the plants that cause internal poisoning. Many of these contain some of the most poisonous materials known to man. Only a few of the more widespread and dangerous plants (figures D-3 through D-11) are included in this manual.

Remember, *if you don't know the plant, don't eat it* unless you have applied the Universal Edibility Test (Chapter 5) and found it safe.

Descriptions of Plants

Figures D-1 through D-11 are pictures and descriptions of a few plants that can cause external poisoning and/or internal poisoning.

Poison ivy and poison oak *(Rhus radicans and Rhus toxicodendron).*

Description: These two woody plants are quite similar in appearance and are often confused. Both have alternate, compound leaves with three leaflets, are deciduous, and have berrylike fruits. The stem may be a rusty brown in color.

Poison ivy is a vine that can grow high in trees. The individual leaflets are only slightly lobed and the grey fruits are not hairy.

Poison oak is often shrubby but can climb. The leaflets of poison oak are usually lobed and resemble oak leaves. The fruits are hairy.

CAUTION: Remember that all parts of poison ivy and poison oak can cause serious dermatitis, especially in sensitive individuals. The fact that various birds and animals feed on these plants does not mean that they are not dangerous for humans.

Habitat and distribution: Look for poison ivy and poison oak in almost any habitat. They are restricted to North America.

Figure D-1. Poison ivy and poison oak.

Poison sumac *(Rhus vernix).*

Description: This is a shrub that grows up to 12 feet tall. It has alternate, pinnately compound leaves; the leaflets are dark green in color and have an entire margin. Flowers are greenish-yellow and inconspicuous. The fruits hang in clusters and are a grey-brown in color.

CAUTION: All parts of the plant should be considered poisonous at all seasons. Contact with this plant can cause severe dermatitis.

Habitat and distribution: Poison sumac grows only in wet acid swamps and is found only in North America.

Figure D-2. Poison sumac.

Castor bean *(Ricinus communis).*

Description: This is a semiwoody plant with large, alternate fan-shaped leaves. The flowers are very small and inconspicuous. The fruits are bome in clusters at the tops of the plants.

CAUTION: All parts of the plants are very poisonous to eat. The seeds are large and may be mistaken for a beanlike food.

Habitat and distribution: This plant is found in all tropical regions.

Figure D-3. Castor bean.

Chinaberry *(Melia azedarach).*

Description: This tree has a spreading crown and grows to a height of 30 feet. It has alternate, compound leaves with toothed leaflets. Its flowers are light purple with a dark center and are bome in ball-like masses. It has marble-size fruit that is light orange when first formed, but lighter in color as it becomes older.

CAUTION: All parts of the tree should be considered dangerous if eaten.

Habitat and distribution: Chinaberry is native to Africa but is now widely planted as an ornamental tree throughout the tropical and subtropical regions. In some areas it has religious significance so may be planted near houses and villages. Otherwise, look for it as a commonly planted street tree.

Uses: The leaves are a natural insecticide and will repel insects from stored fruits and grains. Take care not to ingest leaves mixed with the stored food, however.

The bark can be powdered and used as a fish poison.

Figure D-4. Chinaberry.

Death lily *(Zygadenus).*

Description: This plant arises from a bulb and may be mistaken for an onionlike plant. Its leaves are grasslike. The flowers are six-parted and the petals have a green heart-shaped structure on them. The flowers grow on showy stalks above the leaves.

CAUTION: All parts of this plant are very poisonous.

Habitat and distribution: You will find death lilies in wet, open, sunny habitats although some species favor dry rocky slopes. They are common in parts of the western United States. Some species occur in eastern United States and in parts of western subarctic North America and eastern Siberia.

Figure D-5. Death lily.

Florida poison tree *(Metopium toxiferum).*

Description: This tree grows up to 40 feet high. It has alternate, compound leaves. Each leaf consists of an odd number (3, 5, or 7) of leaflets. The young leaves are dark green and shiny.

CAUTION: This is a dermatitis-causing plant. Contact with it may produce blisters. Smoke from the plant will annoy or injure the eyes.

Habitat and distribution: This tree grows in open, dry woodlands. It is one of the most common trees in the Florida Keys and is found elsewhere in the Caribbean.

Figure D-6. Florida poison tree.

Lantana *(Lantana camara).*

Description: This is a shrublike plant that may grow up to 5 feet high. It has opposite, round leaves and flowers borne in flat-topped clusters. The flower color, which varies in different areas, may be white, yellow, or orange. It has dark-blue or black berrylike fruit. A distinctive feature of all parts of this plant is its strong scent.

CAUTION: All parts of this plant are poisonous if ingested and can be fatal. This plant causes dermatitis in some individuals.

Habitat and distribution: Lantana is grown as an ornamental plant throughout the tropics, but in some areas it has escaped from cultivation and is a weed along roads and in old fields.

Figure D-7. Lantana.

Manchineel *(Hippomane manchineel).*

Description: This is a tree with alternate, shiny green leaves. Its fruits are green or greenish-yellow when ripe.

CAUTION: This tree is extremely toxic. It causes severe dermatitis in most individuals after only one-half hour. The smoke irritates the eyes. No part of this plant should be considered as a source of food.

Habitat and distribution: This tree prefers coastal regions. It is found in south Florida, the Caribbean, Central America, and northern South America.

Figure D-8. Manchineel.

Oleander *(Nerium oleander).*

Description: This shrub has alternate, very straight, dark green leaves. Its flowers may be white, red, pink, or intermediate colors. Its fruit is a brown podlike structure with many small seeds.

CAUTION: All parts of the plant are very poisonous. Do not use the wood for cooking.

Habitat and distribution: This native of the Mediterranean area is now widely planted as an ornamental throughout the tropical and subtropical regions of the world. It is planted in open sunny areas.

Figure D-9. Oleander.

Pimpernel *(Anagallis arvensis).*

Description: This is a small annual plant only a few inches high with opposite leaves. Its flowers are five parted and may be crimson or blue in color. It resembles the edible common chickweed when it is not flowering and could be accidentally collected with that plant.

CAUTION: All parts of the plant are poisonous.

Habitat and distribution: This weed can be found in grain fields, along fields and roads, and in other weedy areas. It is native to Europe but has been introduced into the United States. It is also found in the Middle East and North Africa.

Figure D-10. Pimpernel.

Rosary pea *(Abrus praecatorius).*

Description: This plant is a vine with alternate compound leaves, light purple flowers, and beautiful seeds which are red and black. *One seed may contain enough poison to kill an adult.*

CAUTION: This is one of the most dangerous of all poisonous plants. All parts should be considered poisonous to ingest.

Habitat and distribution: This is a common weed in parts of Africa, southern Florida, and Central and South America.

Figure D-11. Rosary pea.

Index